# Advanced Joining Technologies

# Advanced Joining Technologies

## Proceedings of the International Institute of Welding Congress on Joining Research, July 1990

Edited by
**T. H. North**
*WIC/NSERC Professor,*
*Department of Metallurgy and Materials Science,*
*University of Toronto, Toronto, Canada.*

CHAPMAN AND HALL
LONDON · NEW YORK · TOKYO · MELBOURNE · MADRAS

UK Chapman and Hall, 11 New Fetter Lane, London EC4P 4EE

USA Chapman and Hall, 29 West 35th Street, New York NY10001

JAPAN Chapman and Hall Japan, Thomson Publishing Japan, Hirakawacho Nemoto Building, 7F, 1-7-11 Hirakawa-cho, Chiyoda-ku, Tokyo 102

AUSTRALIA Chapman and Hall Australia, Thomas Nelson Australia, 480 La Trobe Street, PO Box 4725, Melbourne 3000

INDIA Chapman and Hall India, R. Sheshadri, 32 Second Main Road, CIT East, Madras 600 035

---

First edition 1990



Printed in Great Britain by St Edmundsbury Press Limited, Bury St Edmunds, Suffolk

ISBN 0 412 38600 3



**British Library Cataloguing in Publication Data**
International Congress on Joining Research (*1990: Montreal, Canada*)
Advanced joining technologies.
1. Materials. Joining
I. Title II. North, T. H.
670.42

ISBN 0-412-38600-3

**Library of Congress Cataloging-in-Publication Data**
Advanced joining technologies/edited by T. H. North.
p. cm.
ISBN 0-412-38600-3
1. Welded joints. 2. Welding. I. North, T. H., 1943–
TA492.W4A28 1990 90–37893
671.5′2–dc20 CIP

# Contents

# Contributors

**Araya, T.**
Mechanical Engineering Research Laboratory, Hitachi Ltd, Tsuchiura-shi, 300, Japan

**Denys, R. M.**
University of Gent, Gent, Belgium

**Drews, P. and Starke, G.**
University of Aachen, Aachen, West Germany

**Eagar, T. W.**
Department of Materials Science and Engineering, Massachusetts Institute of Technology, Cambridge, Massachusetts, USA

**Easterling, K. E.**
Department of Engineering Materials, University of Luleå, Sweden

**Eichhorn, F.**
University of Aachen, Aachen, West Germany

**Goldak, J. and Bibby, M. J., Downey, D. and Gu, M.**
Mechanical and Aerospace Engineering, Carleton University, Ottawa, Canada

**Heidt, H.**
Bundesanstalt für Materialforschung und -prüfung, Unter den Eichen 87, D-1000 Berlin 45, West Germany

**Johnson, K. I. and Riches, S. T.**
The Welding Institute, Abington Hall, Abington, Cambridge CB1 6AL, UK

**Kuwana, T.**
Department of Materials Processing, Faculty of Engineering, Tohoku University, Sendai 980, Japan

**Lundin, C. D.**
Materials Joining Group, Materials Science and Engineering, University of Tennessee, Knoxville, TN 37996, USA

**Matsuda, F. and Lee, J-B.***
Welding Research Institute, Osaka University, Osaka, Japan (*currently at RIST, Republic of South Korea)

**Nakao, Y., Nishimoto, K., Shinozaki, K. and Kang, C. Y.****
Department of Welding and Production Engineering, Osaka University, Osaka, Japan. **Department of Metallurgical Engineering, Pusan National University, Korea

**Nicholas, M. G.**
Materials Development Division, B552 Harwell Laboratory, UK Atomic Energy Authority, Oxfordshire, OX11 ORA, UK

**Nishiguchi, K.**
Faculty of Engineering, Osaka University, Osaka, Japan

**North, T. H.**
Department of Metallurgy and Materials Science, University of Toronto, Toronto, Canada

**Petershagen, H.**
Institute of Naval Architecture, University of Hamburg, West Germany

**Rothwell, A. B., Dorling, D. V. and Glover, A. G.**
Nova Corporation of Alberta, Alberta Gas Transmission Division, PO Box 2535, Station M, Calgary, Alberta, Canada

**Paton, B. E. and Lapchinskii, V. F.**
E.O. Paton Electric Welding Institute of the Ukr.SSR Academy of Sciences, Kiev, USSR

# Introduction

T. H. NORTH
Dept. of Metallurgy and Materials Science,
University of Toronto.

This volume documents the proceedings of the International Congress on Joining Research held under the auspices of the Canadian Council of the International Institute of Welding in Montreal, July 20/21 1990. This congress was sponsored by the Welding Institute of Canada, Oakville, Ontario.

The study of joining is important, both from the fundamental and applied science points of view. Joining encompasses a wide range of areas, from welding processes, through welding metallurgy and materials science, to non-destructive testing, automation and field construction. Welding has sometimes been referred to as some curious combination of art and science. Certainly, from a university research perspective, the welding area is remarkably difficult to tackle because it is extremely difficult to sift out the critical variables. As a result, it is sometimes difficult to separate the real from the imaginary in any detailed evaluation of the joining literature. I sincerely hope that the authoritative contributions in this volume will sweep away any confusion that exists in the mind of the reader.

As we enter the last decade of the 20th century we are increasingly being informed that our world is changing from one based on a manufacturing economy to one based on an information society. This change in demographics has profound implications for joining specialists since manufacturing and joining go hand-in-hand. Its output is a visible trend from metals to materials' research (to a mix of metals, polymers and ceramics), from conventional arc welding to high-energy joining and solid-state bonding, and from manually-controlled fabrication to the world of automation, robotics, and real-time NDT. If one considers that many present-day welding engineers were trained in Metallurgy Departments (that are now Material Science Departments) and have principally worked on the fusion joining of metals (particularly in steel fabrication) this shift in demographics has disturbing implications. Fortunately, it is likely that tomorrow's world will involve a synergistic mix of materials, rather than a replacement of one material by another. The reader might ask what will become the key emphasis area in the future, and what new skills will tomorrow's welding engineers need to have? Although I am cognizant of the old Chinese proverb, 'to prophesy is difficult, particularly with respect to the future', I feel that this question should be addressed. There is little doubt that fusion welding, and particularly steel fabrication has received the bulk of the research emphasis during the past 30 years, and the great strides in computer modelling (of fusion zone dimensions, thermal and stress/strain cycles, HAZ properties, and so on) are the natural outcome from the mountain of accumulated data. However, it should be borne in mind that in fusion welding of steel we have accepted and largely overcome the heat-affected-zone problem. In future we will be faced with substrates such as composite materials which must be joined with the prime requirement that their intrinsic nature remains unaltered by the joining operation. As a consequence non-fusion joining processes such as brazing, which is very poorly understood at the present time, and solid-state bonding will become increasingly important. If metals, polymers and ceramics are considered as the points of some materials' triangle, the sides of this triangle will likely become the focus for much research. The current interest in ceramic–metal bonding is perhaps the clearest example of this trend. In polymer–metal bonding, the initial joint strength properties depend on adhesion between the metal/metal oxide and metal oxide/polymer layers, and the durability of these joints depends on the stability and long-term properties of the metal oxide layer. In effect, polymer–metal bonding brings together all the points of the materials' triangle. Of course, this assessment is based wholly on a materials' viewpoint. The reader can draw his own conclusions concerning the future trends in other key areas based on a detailed analysis of the information presented in this volume.

The invited papers encompass a comprehensive examination of the joining field. Each presenter was specifically selected based on his reputation in a given topic area, and eight countries were represented (Belgium, Canada, Federal Republic of Germany, Japan, Soviet Union, Sweden, United Kingdom and the United States). I am pleased to indicate that the final presenter list includes many of the foremost experts in the joining field, and may well rate as one of the strongest ever brought together for a single meeting. This formidable presenter list was further augmented by the participation of my Co-Chairmen (Dr Richard Dolby, TWI, Abington and Dr James McGrath, MTL Labs, Ottawa) and four first-rate Session Chairmen, i.e. Dr Bent Koch (Danish Welding Institute, Copenhagen), Dr John Lancaster (United Kingdom), Dr Martin Prager (Executive Director, Welding Research Council, New York) and Professor Konstantin Yuschenko (Deputy Director, Paton Welding Institute, Kiev).

The aim of the Congress involved bringing together a wide spectrum of practising engineers, technologists and research scientists to discuss many aspects of the joining field. It is my hope that the proceedings of this Congress will become an excellent source of information for practising engineers, researchers and students who wish to update their knowledge of welding.

Acknowledgements. This document could not have been published without invaluable advice and support from Mr. Bill Winship of the Welding Institute of Canada. Bill's ever-present vitality, coupled with his long experience of the North American welding and textbook publication industries kept me going through the extended gestation period of this Congress proceedings. Finally, I would like to express my thanks to each presenter for his contribution. It must have been difficult for them to maintain an even keel, bearing in mind the steady rain of communications that came from a Canadian who happens to have an office next door to a telefax machine.

PART ONE

# Research Directions in Joining Processes

F. Eichhorn: "High energy joining"
University of Aachen, West Germany

## A) INTRODUCTION

By comparison with the widespread welding processes which use the electric arc as a source of heat, beam techniques which make use of the electron beam (EB) or the laser beam (LB) have so far only been able to find a limited range of applications in welding engineering, despite their specific advantages and their high level of development. The reasons for this are many and varied. The most important technical and commercial aspects are apparent from Figure 1

**economic** — **characteristics** — **technical**

economic:

- high investment
- low occupation efficiency
- high operating costs

technical:

- beam characteristics
- beam positioning and manipulation
- changing in beam geometry and intensity
- dependancy on material type
- humping effect at high welding speed
- normaly evacuated working chamber required
- unwished beam deflection by magnetic influence at EB
- limited penetration depth

figure 1: essential aspects of the restricted area of application of EB and LB welding

The main feature among the common technical characteristics is the small beam cross section with particularly high power density which is highly desirable on the one hand as regards the single-pass technique, the concentration of the thermal power on the boundary surfaces of the two parts to be joined which makes possible a high degree of thermal efficiency, a high welding speed and a very low distortion of the component during cooling. Even the generally unfavourable thermal influencing of structure and properties is minimised in this way.

On the other hand, however, the small effective beam diameter calls for very precise preparation of the parts to be joined and alignment and maintenance of these parts without any gap - or at the most with only a very slight gap - if the fundamental advantage of being able to work without using filling materials is to be retained. Moreover the beam must be directed at the joint groove with a high degree of precision and be able to follow it exactly if welding defects are to be avoided.

Despite this during the past years substantial progress has been achieved in both fields of beam techniques, electron beam welding (EB welding) and the younger process of laser beam welding (LB welding) which is still right in the middle of the development stage, so that in the near future an expanded range of applications can be expected. The general trend towards increasing mechanization in welding engineering supports this tendency, since fully mechanized systems were used here right from the start. The good controllability of the beam geometry and the distribution of intensity as well as the total power of the beam as a function of time and the travel control of the beam allow a flexible adaptation to the specific requirements in each case.

Alterations of the beam at constant setting parameters can be identified in good time by diagnostic beam monitoring and can be compensated before they affect the welding result to an impermissible extent.

Despite many factors in common there is a series of specific features of the two systems with very high power density and at the same time a small cross section action which is shown in Figure 2 and which opens up differing fields of application for these systems alongside the competition for certain joining operations in the thin plate section.

| laser beam | electron beam |
|---|---|
| - good handling | - normally evacuated work chamber |
| - limited penetration depth | - enormous workingpiece motion |
| - totally efficiency low | - beam controllability free of inetria |
| - shielding gas necessary | - protection against X-ray |

figure 2: special features of EB welding in compairison with LB welding

## B) STATUS AND DEVELOPMENT TRENDS IN EB WELDING

The developments of the past years, especially in the fields of improvement of the reliability of the installations by improved process analysis and process control, composite and adaptive beam guidance and process management systems have now reached a stage which allows a high degree of quality assurance during the welding process and an excellent transferability and reproducibility of the work results insofar as the variations of the material composition move within specified limits. Signals which are suitable for an adaptive process control are shown schematically in Figure 3. Of these those which are most suitable for manufacturing engineering applications are the backscattered electron flow and the emission of X-ray radiation for evaluation in adaptive systems for tracking the electron beam along the welding groove. The penetration flow can be used for through-welding operations.

In the case of closed seams, e.g. round seams on cylindrical containers, special attention must be paid to defect-free overlapping of the beginning and end of the seam. Defect-free closed seams (Figure 4) can be produced by a suitable program for modifying the beam parameters in the area of the overlap. An example for adaptive tracking of the EB along the groove is shown in Figure 5. Here the EB itself is used even at full power as a measuring tool for raster scanning of the welding groove in order to keep the beam positioned in the middle of the groove as the welding progresses with the aid of the backscattered electrons by means of an appropriate movement of the workpiece. Since the measuring operation takes place at a high scanning speed of the beam (500 to 900 m/s) in about 1 ms, the workpiece is not thermally damaged and no welding defects occur as a result of brief interruption of the welding process. By co-turning of the measuring field with the workpiece movement, curved seams can also be produced in this way without having to be programmed.

On the grounds of the high weld freezing and cooling rate of the small molten pool in conjunction with a high temperature gradient, special demands are made of the steel materials as regards hot-crack promoting companion elements and the gas contents. Hot cracks as illustrated in Figure 6 can be avoided in part by technological measures, in particular by enlarging the molten pool and at the same time reducing the welding speed (Figure 7) provided that with a horizontal beam arrangement and transverse weaving scanning a beam-power range is maintained which still allows a uniform bead formation on the other side, but which at the same time does not lead to the molten pool

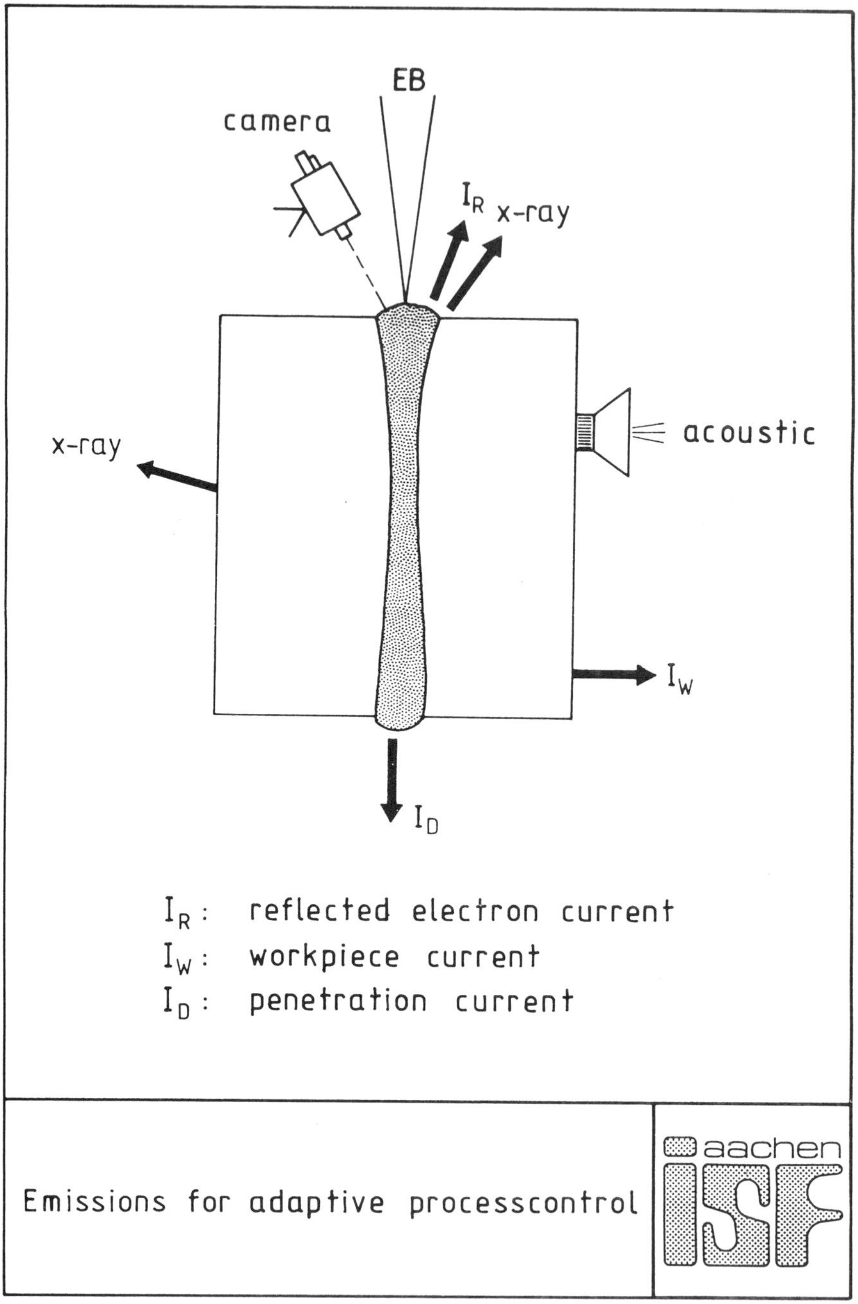


*Fig 3* Emissions for adaptive process control

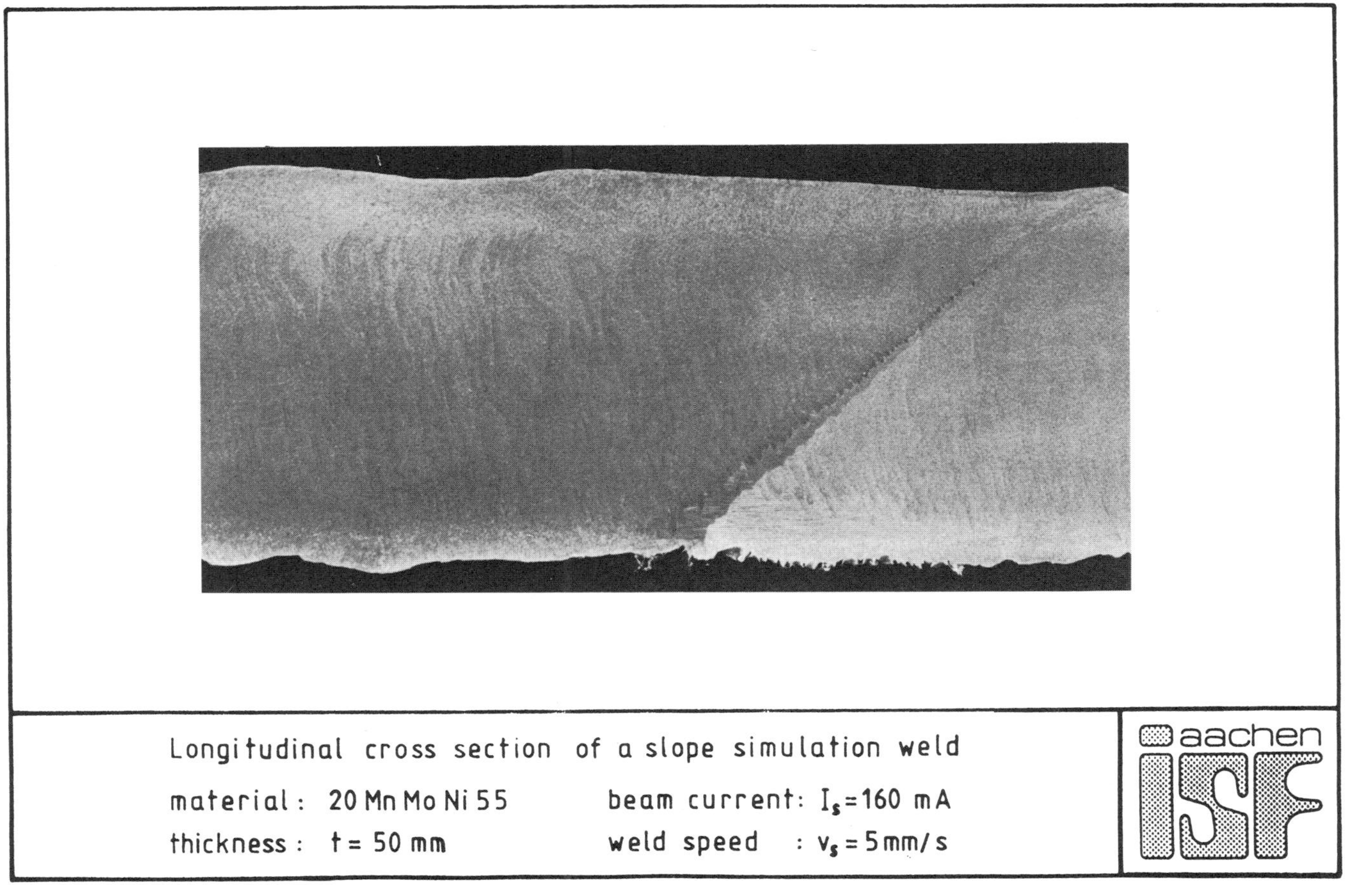


*Fig 4* Longitudinal cross section of a slope simulation weld

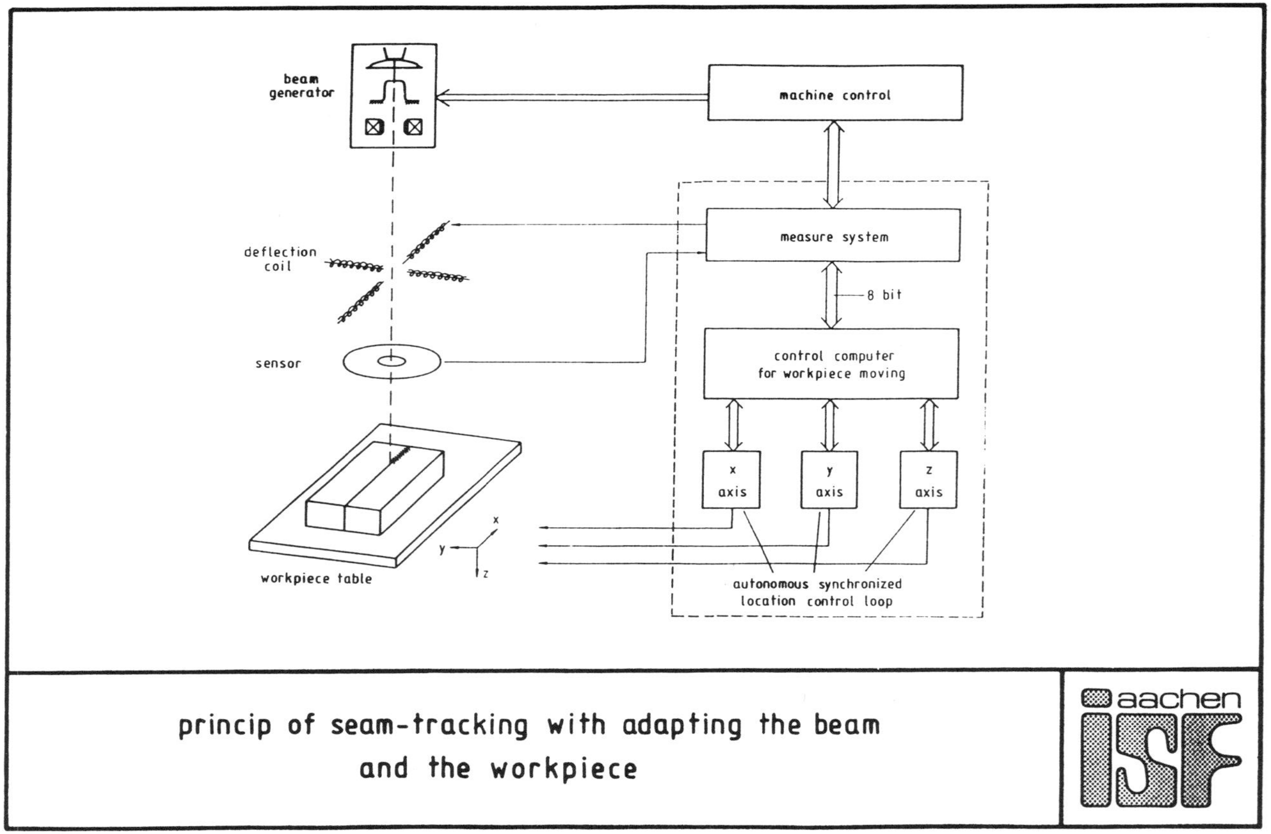


*Fig 5* Principle of seam-tracking with adapting the beam and the workpiece

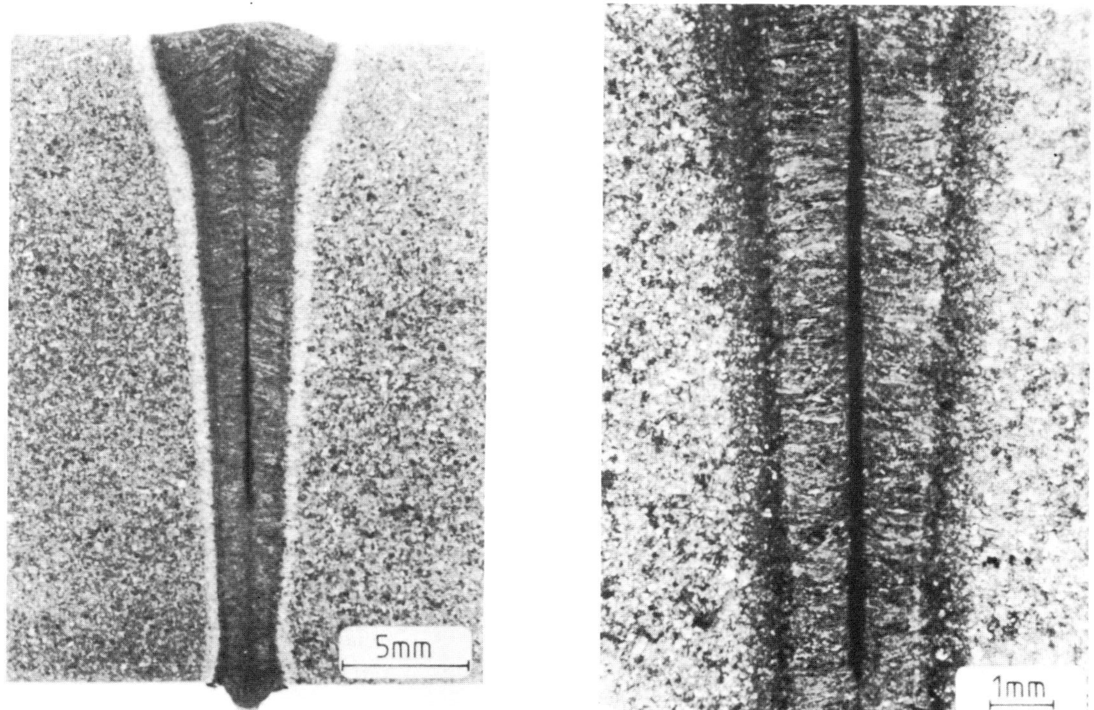


*Fig 6* Macro-crack in the middle of the melting zone of an EB-welding

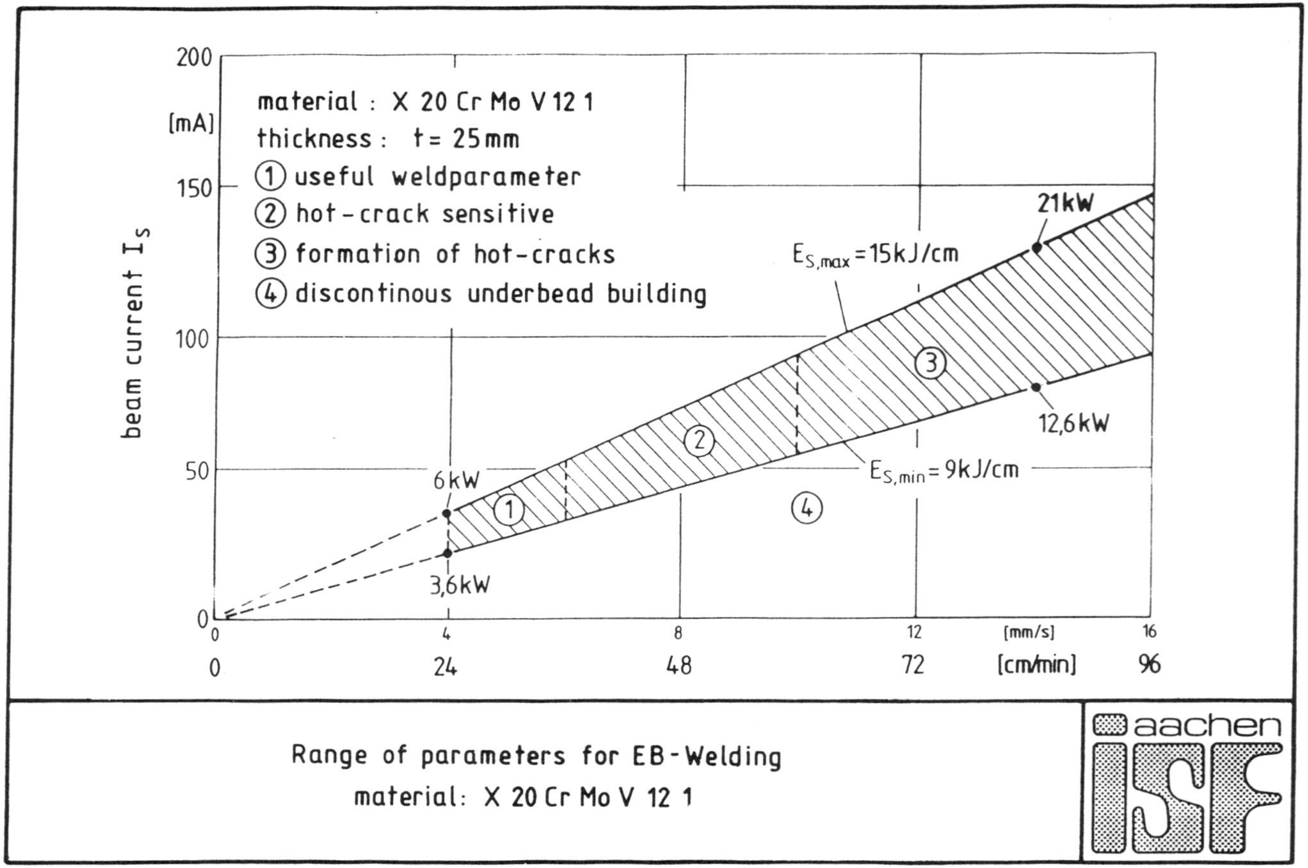


*Fig 7* Range of parameters for EB-welding

flowing out at the workpiece surfaces at raised power levels.

The influence of the gas contents on thick steel plates is shown in Figure 8, which makes it evident that in particular the nitrogen content should be limited to about 100 ppm for the creep-resistant steel under review here with wall thicknesses of 100 mm insofar as the usual chamber pressures of about 1 Pa are used.

Today's computer-supported and easy to use description of the thermal tool "electron beam" in relation to the beam form (aperture) and power density distribution over the cross section allows uniform quality to be assured over long production periods and provides warnings in good time e.g. of the necessity for a cathode change in the event of an altered beam configuration in an EB installation. On the other hand, however, it also offers the possibility of reproducibility and trans-ferability of welding results to other welding machines. This means that in future the input required for working out favourable welding parameters for a certain welding task is reduced and in this way the scope of application and the production security are increased. The use of on-line seam-tracking systems and the automatic through-welding rule as well as an automatic focus adjustment thus allow adapted welding of workpieces of varying thickness and variable working distance.

At the present time large-chamber installations with high power EB facilities are in preparation in both the UK and France for thick-plate welding in order to utilise the technological advantages of welding in a single operation with minimal distortion and minimised thermal influencing of the structure at high welding speeds.

However the electron beam also has great potential for applications in the field of delicate and miniature welding with small cross sections thanks to its good inertiafree controllability and small active cross section. In this field, however, there is growing competition from the laser beam which offers certain advantages, aspecially as regards simpler handling without the necessity for a working chamber.

It remains to be seen to what extent welding under atmospheric pressure without a working chamber as is already practiced in the USA in particular will prevail in competition with electric arc or plasma arc welding or LB welding. A hindrance here alongside the elevated outlay for the constant evacuation of the locks is the divergence of the EB steel which leads to a trouble-prone short working interval of the welding equipment with restricted

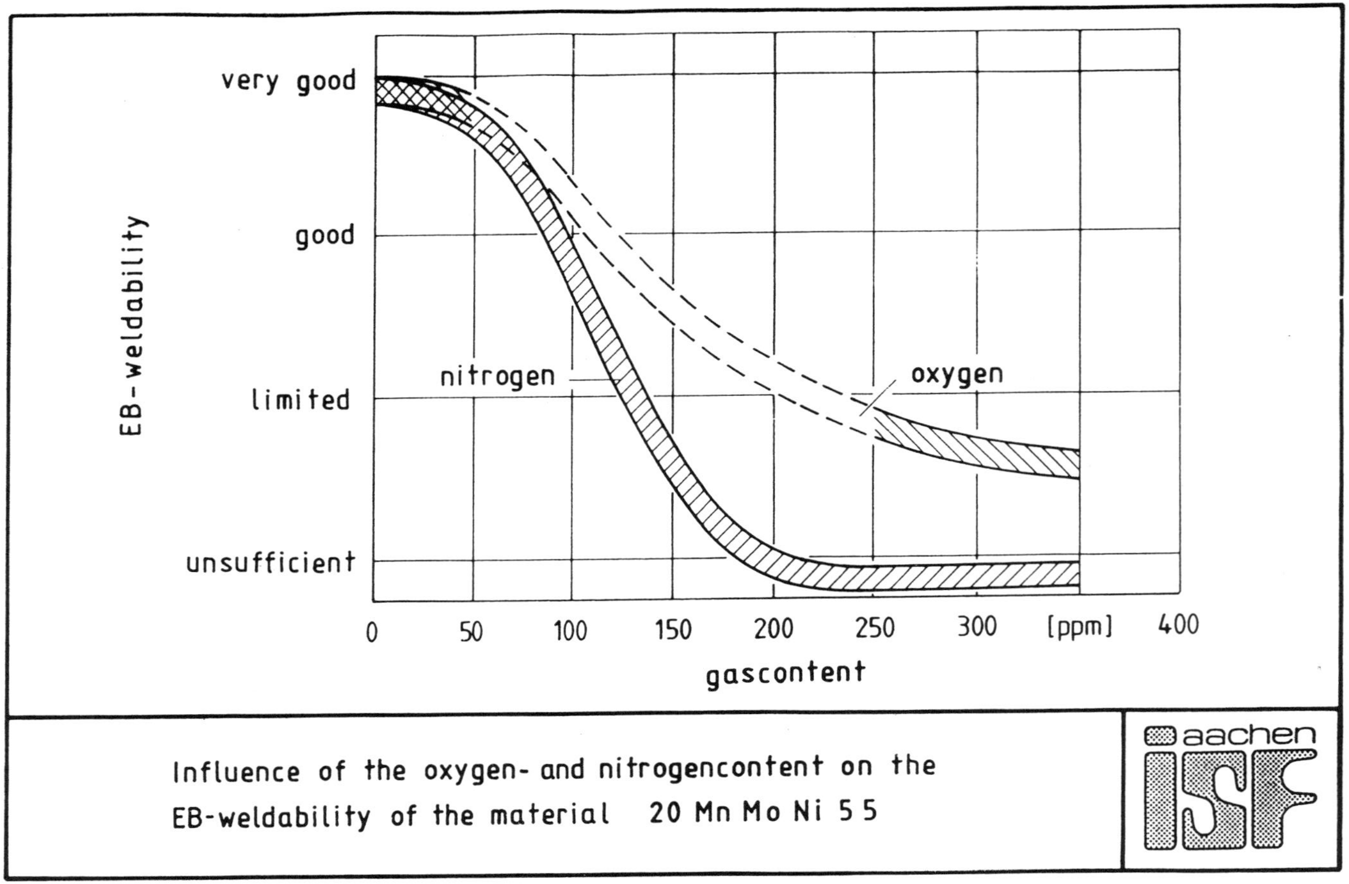


*Fig 8* Influence of the oxygen – and nitrogen – content on the EB-weldability of the material 20MnMoNi55

possibilities for observing the welding point and produces a welded joint which is similar to a plasma seam. However this does not exhaust a part of the possible advantages of the high power beam technique and it is generally not possible to do without the use of a shielding gas at the welding point.

## C) STATUS AND DEVELOPMENT TRENDS OF LB WELDING

In the still young field of LB welding a development is currently underway which has been largely concluded now for EB welding in connection with the beam analysis, process analysis with the interaction between beam and material, process control and adaptive process management. Two examples from the Fraunhofer Institut für Lasertechnik (Institute for Laser Technology) in Aachen are intended to show how the weld sink of the top bead through changes in gap width can be detected with the aid of signals from the lateral observation of the intensity of the plasma at the component surface (Figure 9). By visual observation of the underbead too changes in the penetration of the component, e.g. in the case of changes in wall thickness or fluctuations in intensity of the beam, can be identified (Figure 10).

More recent development avenues of the lasers used for material processing, which account for the greater part of the laser market, aim not only at controlling and regulating the laser on the basis of signals from the observation zone, but also at improved and more stable beam qualities by means of new kinds of resonator structures and modular design for gas laser powers of more than 10 kW. At powers of up to about 10 kW the axially streamed $CO_2$ laser appears to have the edge over the transversely streamed laser on the grounds of its better beam quality, although with its less compact structure it can evidently not be produced so cheaply. Moreover stimulation by high-frequency fields appears to be gaining acceptance since this allows higher power densities to be achieved. Favourable impacts on the welding result can also be achieved by making use of polarisation effects.

Alongside the $CO_2$ observation laser for higher welding powers, however, the solid-state type laser is also becoming more important, not only for delicate and miniature welding now that it allows power enhancements of up to more than 1 kW and improvements of the poor efficiency degree to more than 1 - 2 %. However such systems with a modular structure and with individual modules of 300-400 W generally have a poorer beam quality. An attractive feature here is the possibility of flexible transport of the beam energy via optical fibers thanks to the small wavelength of 1.6 µm, although here coupling in

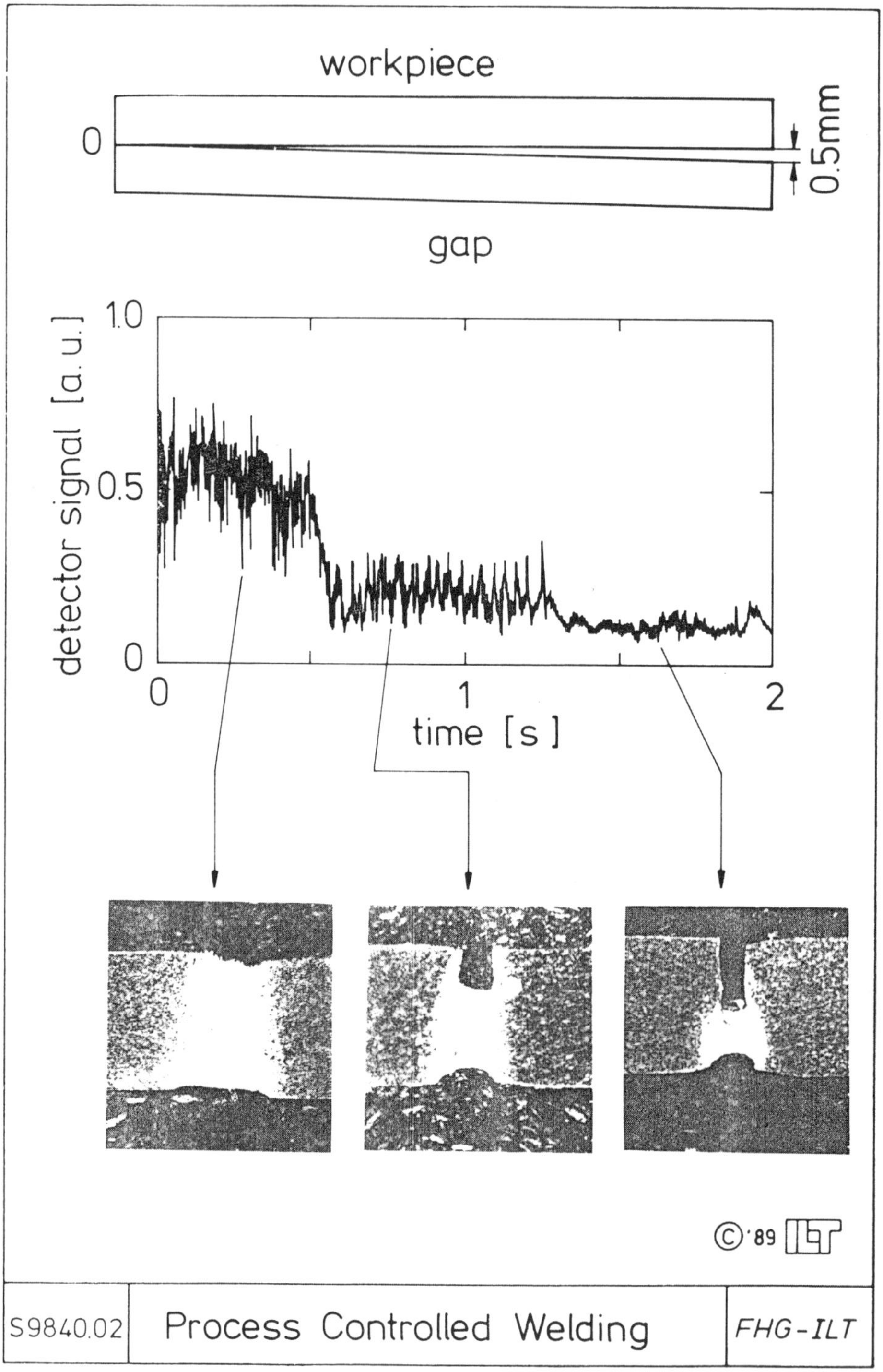


*Fig 9* Process controlled welding

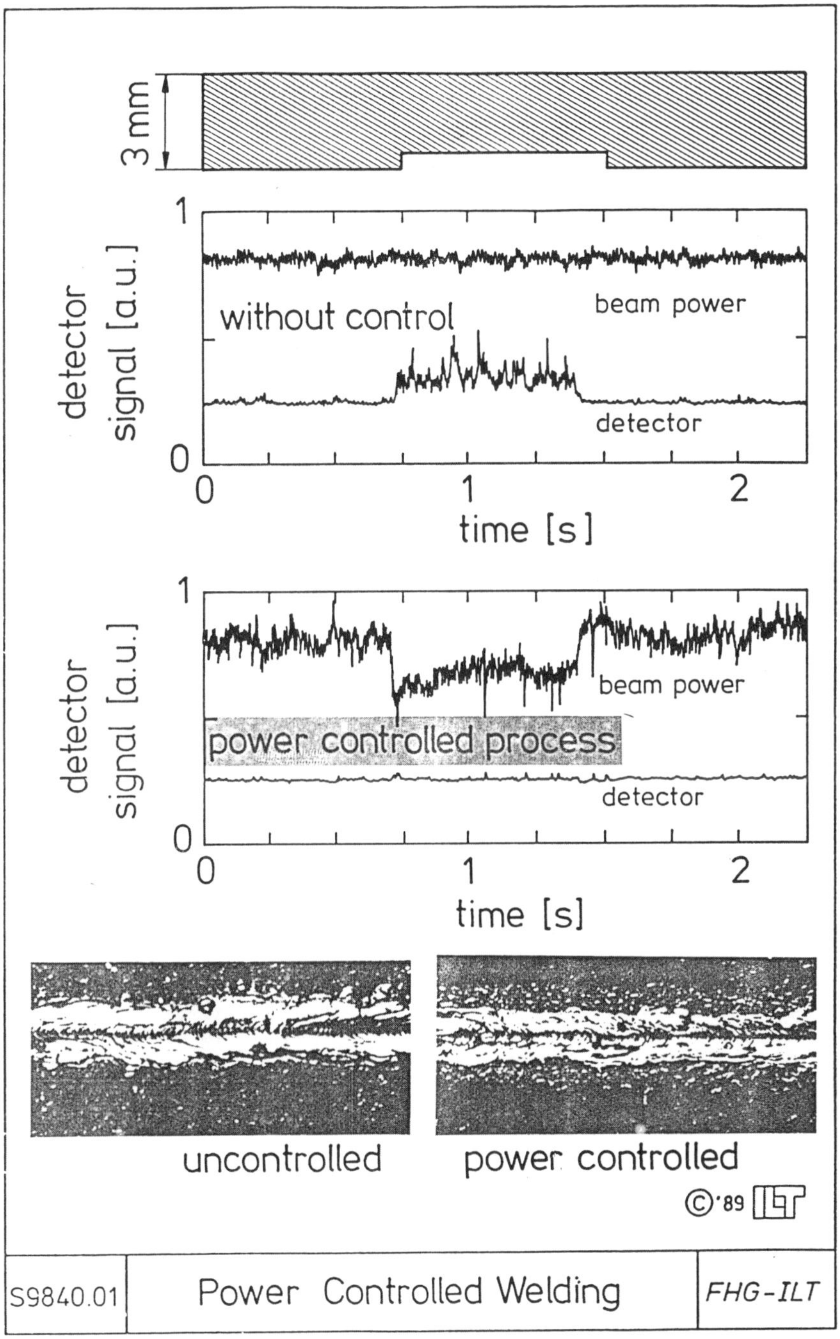


*Fig 10* Power controlled welding

problems and the often unsatisfactory beam quality have to be overcome. Then flexible laser robots could experience a special development impetus with this. Laser diodes are being tried out as a pumping source. Further possibilities of using HF-excited gas discharges as a pumping source are still at their initial phase. In the USA the development goal is a solid-state type laser with a mean power of 200 W at an efficiency of 10 % and a service life of 4 years.

Altogether today's lasers do not yet reach what appears to be the optimum level available for processing tasks. In the medium term, however, adaptation to the practical needs in processing technology, especially in welding technology, is expected. Even today material processing accounts for the largest share of the market in laser production and shows a rising trend. On a worldwide basis an increase of 14 % over the preceding year was noted in 1989, whereby turnover in the beam sources is divided at the rate of approx. 2/3 for $CO_2$ lasers and 1/3 for solid-state type lasers.

One of the principal advantages of laser welding over electron beam welding is considered to be the better handling capability of the beam since no working chamber is used and there is no necessity for screening against X-ray radiation, while at the same time good utilisation of a beam source is achieved by directing the beam to different welding workplaces.

Thus the duty cycle for the beam can be kept high and the secondary times in which no welding takes place can be kept low. However, a shielding gas must generally be supplied to the welding place to screen off the atmosphere not only for gas-sensitive materials, but also for steels.

By making use of the deep-welding effect wall thicknesses of up to about 25 mm in steels can still be joined at justifiable welding speeds with high power $CO_2$ lasers (Figure 11). For production line applications, however, the useful welding- in depths of the LB system are substantially smaller than those of the EB system for both commercial and physical-technical reasons. For thin plates up to a mean steel thickness range of 5 to 8 mm, the advantages of laser welding often prevail. For larger thicknesses under certain circumstances a com-bination of LB welding with an electric arc process (Figure 12) can be considered if both types of welding installation are available. To start with, at the closed square butt joint without any filler material the parts to be joined are aligned and maintained against one another in a cross section which can be selected larger than a normal tack weld and which largely prevents subsequent angle shrinkage. This is followed by a single-sided or double-sided filling

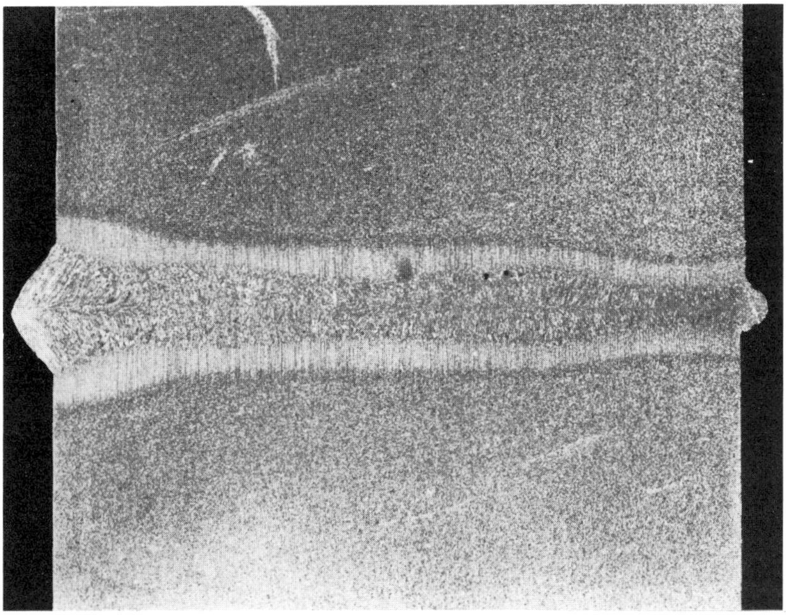

2 mm

horizontal beam | $v_s$ = 0,9 m/min
material: St 52-3 | $P_L$ = 20 kW
He = 15 l/min | t = 20 mm

Cross section through the LB-welding with horizontal beam on a structural steel

aachen ISF

*Fig 11* Cross-section through the LB-welding with horizontal beam on a structural steel

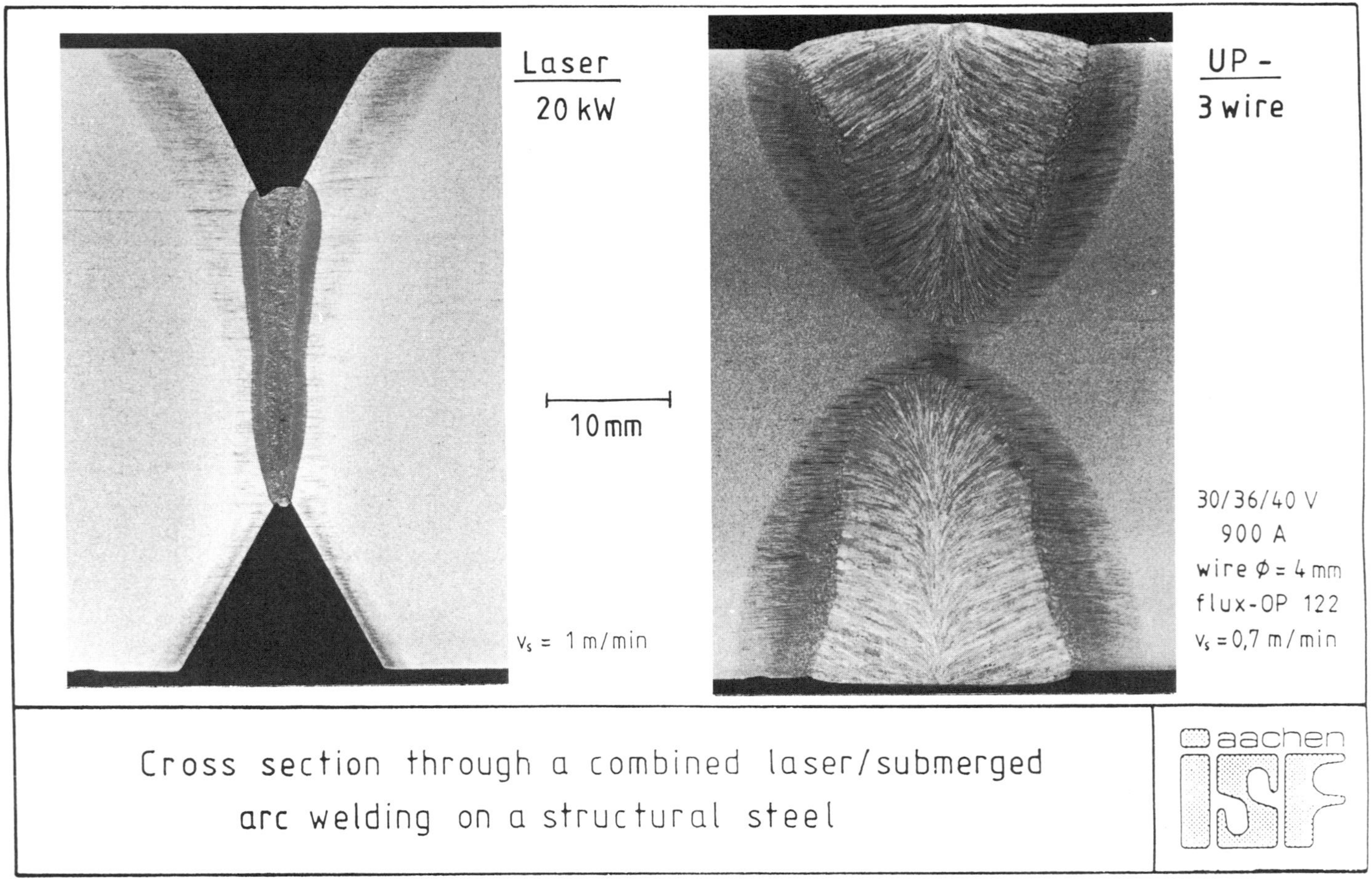


*Fig 12* Cross-section through a combined laser/submerged arc welding on a structural steel

of the open groove with a powerful electric arc process. Since good root fusion is achieved in this way, in the case of one-sided filler pass welding it is generally possible to do without any root rewelding on the opposite side. In the case of double-sided filler pass welding (Figure 12) the otherwise usual remelting of the tack weld or the root of the back-up weld can be dispensed with. Thus in addition to the quality advantages, by means of a lower thermal input and a good root fusion the volume of filling material and the time required for the welding operation can be reduced substantially.

When welding a butt-type joint the high sensitivity of the beam process to an enlargement in the gap between the parts to be joined must be noted. In the same way poor alignment of the plate edges also creates difficulties (Figure 13). In such cases it can be expedient to work with filling materials (Figure 14). However a careful geometrical and volumetric coordination of the filler material feed and the local features as well as the laser beam is necessary. In the event of altered groove geometries during a welding, adaptive adjustment must be made with a suitable sensory system, e.g. as shown in (Figure 15).

In the case of laser weldings of overlapping plates a changed gap between the plates can lead to weld defects if this is not minimized by suitable clamping devices. It may be necessary to work with filling materials here too.

However the LB method can be used not merely for fusion welding but also for pressure welding. One example for this is the longitudinal seam welding of steel pipe made from strip-form material. Here the laser beam is introduced into the groove of the slotted pipe from the front and focussed by wall reflections on the contact point at which the welding of the preheated and liquefied material takes place at the joint place by pressure rollers.

Thermal coating coating of a powdery surfacing material feed is also possible with the laser beam. Finally, the laser system can also be used for producing spot joints on overlapping thin plates. It is not yet possible to foretell to what extent this can replace the cheaper resistance spot welding in individual cases.

Altogether the laser has developed to become an interesting thermal beam tool suitable for a variety of applications and the range of use for joining in particular thin plates by welding is growing constantly.

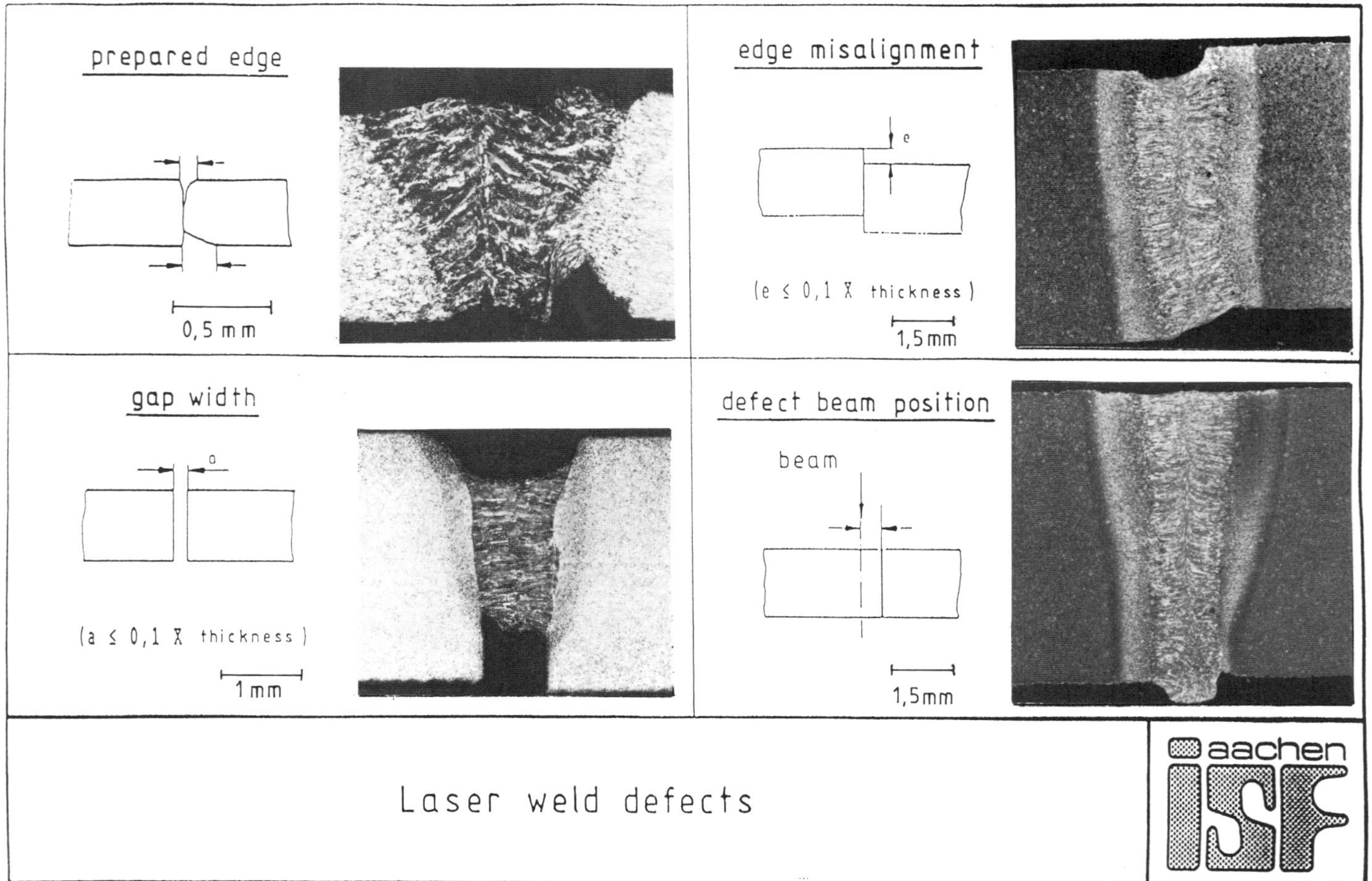


*Fig 13* Laser weld defects

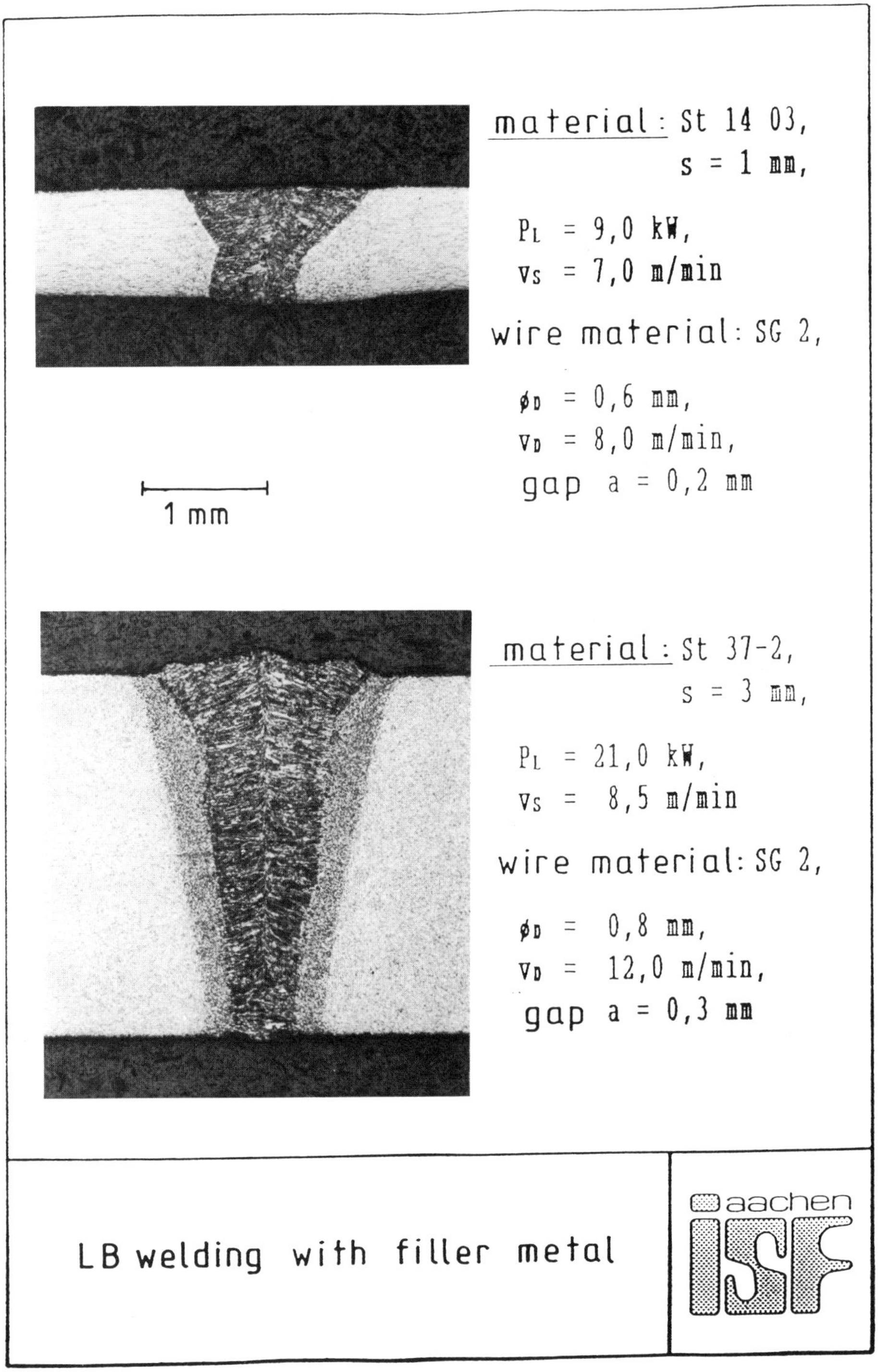


*Fig 14* LB welding with filler metal

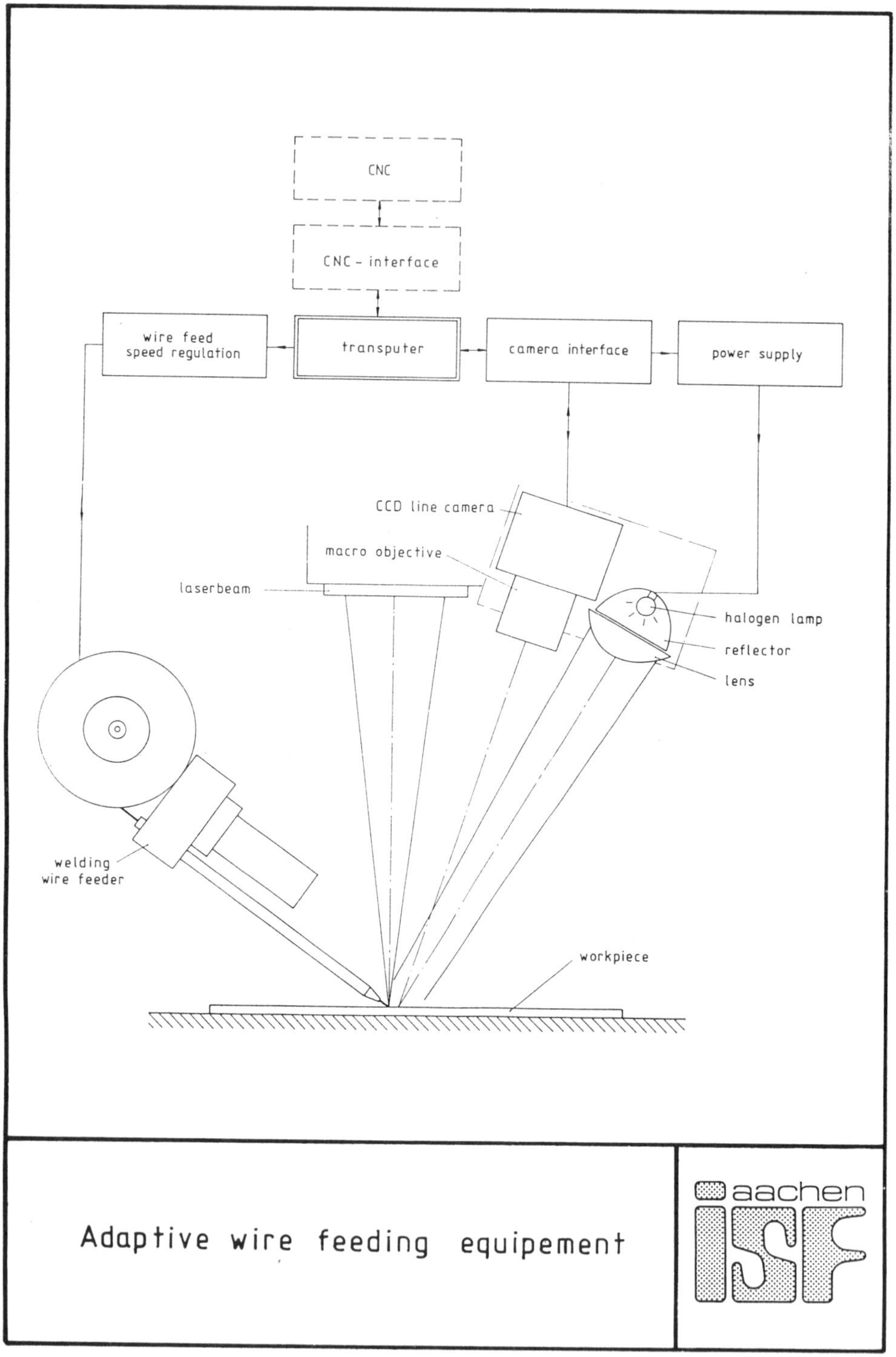


*Fig 15* Adaptive wire feeding equipment

# CURRENT AND FUTURE TRENDS IN MICROJOINING

K I JOHNSON AND S T RICHES

Welding Institute, UK

Abstract
Major developments in microjoining applications for the electronics and miniature fabrication industries are remained including: attachment of active circuits, interconnection of circuits, packaging, package interconnection and transducer and device fabrication. The joining processes employed in the drive to increased miniaturisation, productivity and reliability include adhesives, Si/glass bonding, electrostatic bonding, ultrasonic welding, soldering and miniature arc, resistance and Nd-YAG laser welding.

## 1. Introduction

The involvement of microjoining in the electronics and instrument manufacturing industries is as important and as extensive as that of welding in other manufacturing industries. The proper functioning of small mechanical devices and electronic control systems, as employed in the aerospace, medical, telecommunications, computer, automotive, and consumer industries, is crucially dependant upon the use of efficient and reliable joining techniques.

The major microjoining operatives in the microelectronics industry involve:

- attachment of active devices to substrate circuitry (Si die and wafer bonding),
- interconnection of active devices (chips),
- device/substrate packaging,
- package interconnection.

The instrument manufacturing and electronics industries are also involved in a number of microjoining or miniature welding operations for the manufacture of transducers, gauges optoelectronic devices, etc.

This paper reviews current microjoining applications and future developments in these industries.

Attachment of Active Devices

Active silicon and other circuits and passive resistors, capacitors, etc, have to be attached to package bases or to hybrid circuitry. Four techniques are available for die bonding: gold-silicon eutectic bonding, adhesive bonding, solder, and other inorganics (Ag/glass). The typical arrangements is shown in Fig 1.

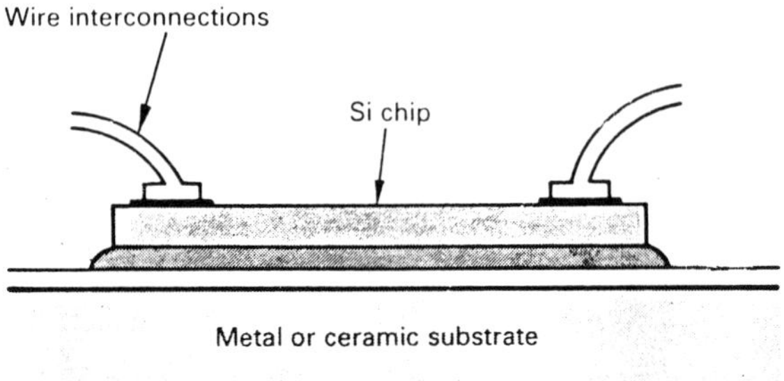


Fig 1. Typical arrangement for die bonding.

Very large scale integration (VLSI) and the increased power of Si chips has resulted in growth of chip size and an increase in the need for power dissipation from the chip. Single chips of up to 15 x 15mm and of 3W power are used currently, and further increases in chip size (50mm x 50mm for Si substrates) and power dissipation (40W for some bipdar devices) will place additional requirements on the die/substrate bond.

A comparison of die attach materials has recently been completed for solder, thermoset adhesives, thermoplastics and Ag/glass large area die (10 x 10)mm Si die attached to $Al_2O_3$ substrates. The selection of a die attached material for a particular application is not straight forward as ability to withstand thermal expansion mismatch and subsequent processing requirements, thermal resistance, stability and device degradation cannot always be predicted from bulk material data. Furthermore, desirable characteristics are often conflicting. Low thermal resistance can be achieved by solders and Ag/glass but solders usually require fluxes and Ag/glass materials require processing (>400°C) to achieve satisfactory bonds. Thermoset epoxy based adhesives are attractive, low temperature materials and low modulus systems have been developed to reduce die cracking, but at the expense of high temperature stability (above 250°C). For polyimide paste adhesive, the removal of solvent is the most critical factor affecting bond quality.

Film/ribbon adhesives have advantages in applying a uniform amount of adhesive over large die areas. Film adhesives require only minor modifications to die bonding parameters but are cured off-line.

Interest in thermoplastic ribbon adhesives is growing, since no off-line curing is required and rework is possible. However, the method of heating is critical if porosity is to be controlled.

For multichip modules, the benefits of using a Si substrate are being explored where advantages in substrate thermal conductivity, elimination of chip/substrate thermal expansion mismatch problems, compactness and lower costs are being found. The Si chip/substrate interconnection can be carried out by wire bonding, TAB or flip-chip. The Si substrates (50mm x 50mm) are attached to the package by film adhesive.

Si/Si and Si/glass bonding is also of growing importance in the manufacture of increasingly sophisticated transducers and sensors. The techniques listed above can be employed, but direct electrostatic bonding is attractive. Components are heated to a temperature below the softening point of glass (300-500°C), loaded lightly and a DC voltage of 3KV applied. Bonding occurs because of electrostatic forces generated by ion migration. The process has been used for joining Si to glass in the manufacture of solar cells and pressure transducers (Fig 2). More recently, Si wafers up to 75mm diameter have been bonded to glass discs for the manufacturer of a prototype sensor arrays (Fig 3).

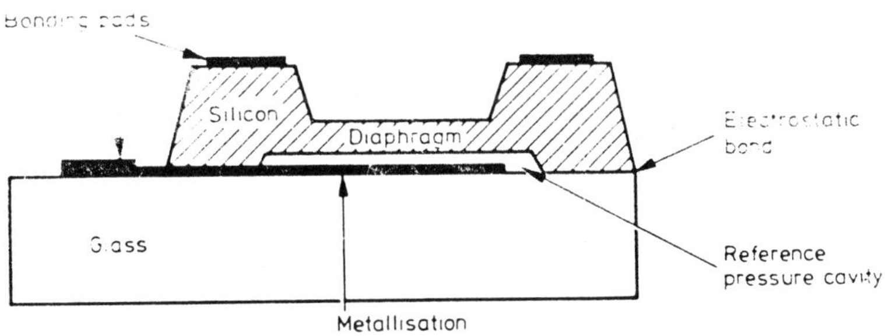


Fig 2. Schematic arrangement of electrostatically bonded Si capacitive transducers.

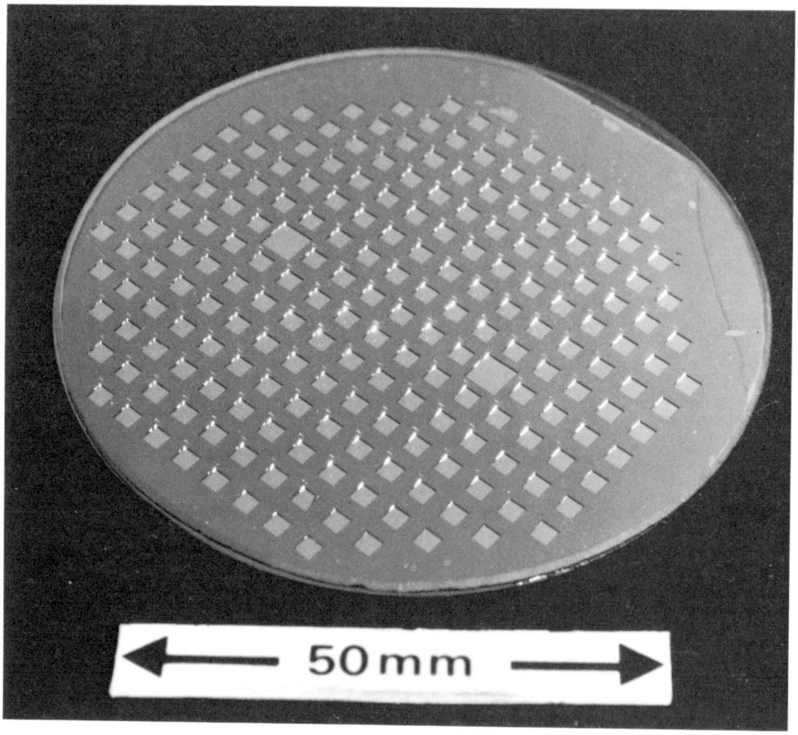


Fig 3. Electrostatically bonded 75 mm diameter etched Si wafer/glass joint for pressure transducer application.

## Interconnection of Active Devices

The most widely used method of incorporating transistor or silicon integrated circuits into more complex circuits is to die bond the reverse side of the devices to the substrate or package base, as just described. The aluminum metallised pads on the surface of the chip are connected to the circuit track on the substrate or to the external leads of the package by aluminum or gold wires (typically 25μm diameter wire for low power signal devices), Fig 4. These wire/pad and wire/circuit tracks or wire/lead joints are made by ultrasonic, thermo- compression (hot pressure), or thermosonic (a combination of ultrasonic and thermocompression) welding. Wire bonding is the most common technique for making interconnections to solid state circuit chips both because it is flexible and suitable for nonstandard chips, and because intensive development by circuit manufacturers and machine suppliers has led to high production rates and high reliability using automatic wire bonding equipment.

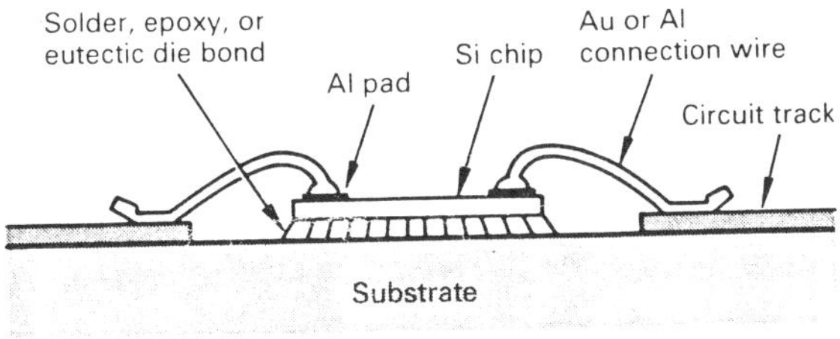


Fig 4. Principle of coventional die and wire interconnection of active devices.

Cu wire bonding has also been developed for device interconnection, Fig 5, and is increasingly finding applications as a cheaper alternative to Au wire.

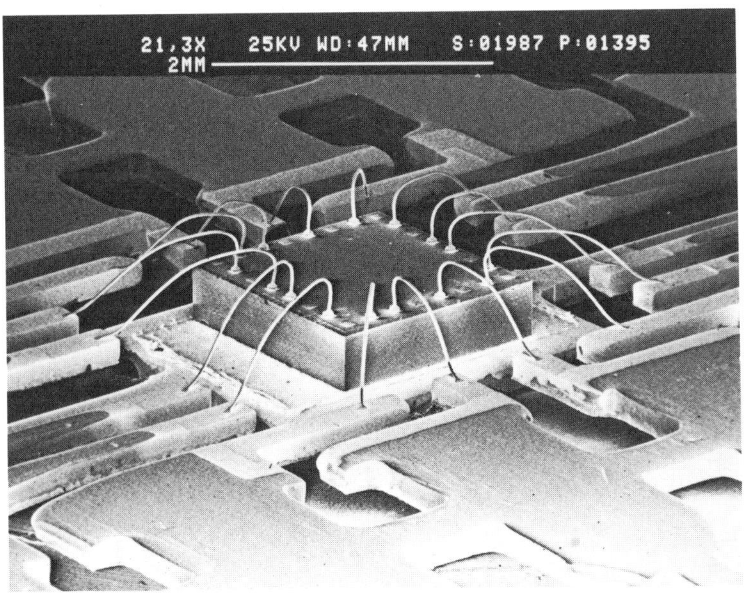


Fig 5. Integrated circuit ultrasonically ball/wedge bonded using 25 µm diameter Cu wire.

The need to connect VLSI type devices having 500 contacts and more has resulted in pressure to reduce bond pad size to 50µm and below. However, it is difficult to use coventional wire bonding techniques on small pad/high pin count devices and work to refine it has been undertaken.

The feasibility of ultrasonically bonding 7µm diameter Au wire has been investigated with modified wedge tool and bonding equipment. Although bonds have been achieved, high wire deformation was encountered and further machine develop- ment is required to enable consistent achievement of bonding forces down to 40mN.

Fine pitch bonding gives an increased likelihood of short circuits between connecting wires. The development of insulation coated wires should alleviate this problem and permit further reduction in bond pitch. The feasibility of ultrasonic bonding anodised Al based wire and polymer insulated Cu wire without pretreatment has been demonstrated.

In Japan, North America and Europe research and development into Tape Automated Bonding (TAB), flip chip, thin film interconnection and polymer bonding techniques has progressed because they offer increased miniaturisation, reliability and performance. Fig 6(a) illustrates the principle and 6(b) a device developed on a European ESPRIT project. The device has 284 leads on a 125µm lead pitch.

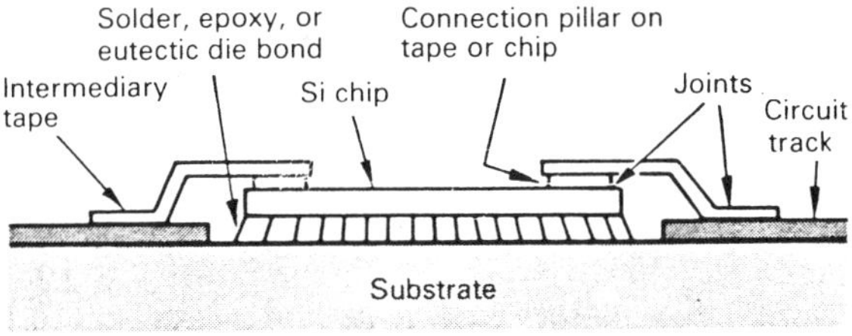


(a) principle of technique

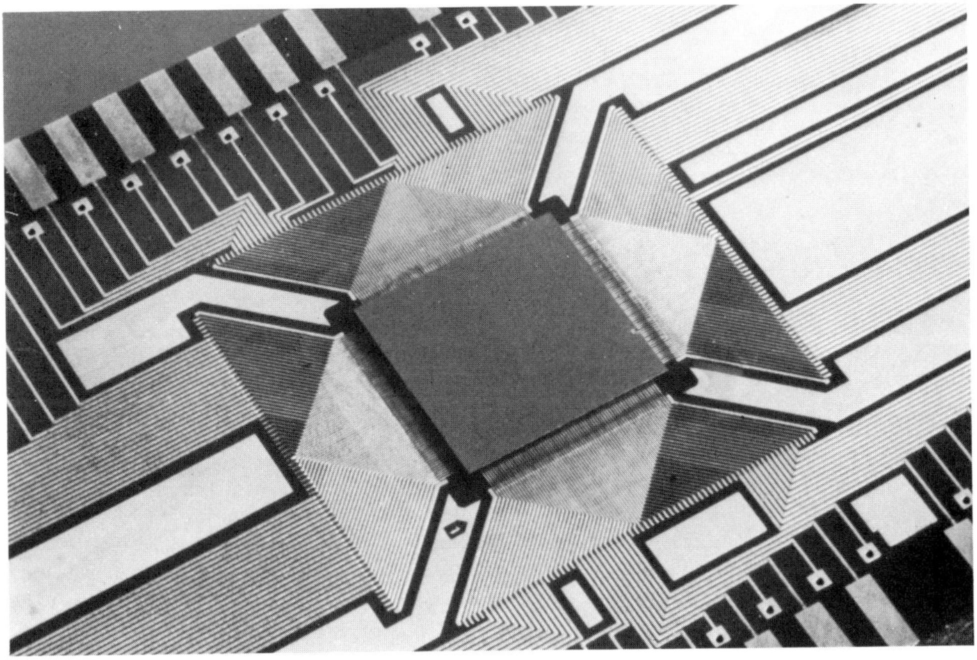

(b) 284 Lead (125µm pitch) TAB device bonded to test board (Courtesy of Bull, France, GEC Marconi and British Telecom, UK, developed on European ESPRIT 958 project)

Fig 6. Tape automated Bonding (TAB)

An alternative technique which combines the performance and fine feature size of TAB bonding and the flexibility of wire bonding is ribbon bonding. It has potential for the interconnection of high lead count, high density VLSI devices or high speed, high frequency microwave/opto-electronic devices, particularly in small batch production, Fig 7.

Fig 7. Device ultrasonically bonded with 30µm x 8µm Au ribbon (Welding Institute, UK)

Device/Substrate Packaging

Electronic circuits are normally protected by plastic packages, except where service conditions are particularly severe, or the device especially sophisticated, in which case metal or ceramic packages are employed. The plastics packaging of electronic devices has developed along two main routes: transfer injection moulding has been used for integrated circuit packaging while encapsulating resins have been used mainly to protect chips on circuit board substrates (chip on board) or TAB devices. However, both techniques can give rise to problems in the packaging of advanced electronic devices where the increased chip size and reduction in package size may cause problems with device functionality, cracking of the package/resin and failure due to moisture penetration.

The enclosure of integrated circuits by thermoplastic lids and bases or the localised lidding of areas of circuit boards has inherent advantages over transfer injection moulding and encapsulation (glob topping) as there is no contact between the plastics encapsulation and the Si chip or interconnect system. This eliminates the problems caused by stresses on the die, interconnection or substrate due to the encapsulant.

The feasibility of ultrasonically welding thermoplastic materials (eg polyarylimide) to encapsulate leadframes (Ni–Fe–Co) and to various substrates (epoxy/glass and polypolymide printed circuit board, polyethermide substrates and polyimide tape) has been demonstrated, Fig 8.

Fig 8. Ultrasonically welded polyarylimide thermoplastic lid to epoxy/glass printed circuit board for device encapsulation (Welding Institute, UK)

Further work is planned to establish the properties of joints between the thermoplastic lids and particularly the circuit board substrates to determine whether the technique can be applied to electronic packaging.

## Package Interconnection

The reduction in component size and the increasing complexity of devices has resulted in an increase in package pin count an closer pitches on circuit board substrates. This trend applies to conventional ceramic and plastic packages, as well as to TAB devices. For these components up to 500 leads at pitches of down to 100μm are being quoted. Such pitches cannot be met with solder deposition by screen or stencil printing (400μm minimum lead pitch). The more precise method of plating is employed where consistent solder plated conductor tracks of down to 60μm width for outer lead bonding flip TAB devices have been produced, Fig 9.

Mass soldering techniques such as vapour phase or infra-red reflow give inadequate control for such fine geometries. The favoured technique for outer lead bonding is pulse heated hot bar reflow, but there is interest in examining single point soldering techniques such as resistance parallel gap or laser soldering to overcome some of the difficulties encountered with ensuring the consistency of soldering up to 500 leads simultaneously.

An alternative to the solder connection of surface mount packages is that of Ag loaded adhesives. This approach is attractive because adhesives are easy to apply, give low processing costs, do not require preplated components, have low curing temperatures and are suitable for fine pitch leads. Thermoplastic materials may also enable rework. Future work is planned to evaluate this approach for the connection of active and passive components on surface mount boards and ceramics substrates. Techniques for the inspection of such joints will also be investigated.

Fig 9. Outer lead bonding of flip TAB device (284 leads) to solder plated conductor tracks (65μm width) on printed circuit board(Courtesy of Bull, France, GEC Marconi and British Telecom, UK developed on European ESPRIT 958 project)

### Miniature Welding

As well as the electronics industry, the small instrument and transducer manufacturing industries have developed increas- ingly sophisticated, complex miniature devices which have required an extension of the capabilities of the joining techniques employed.

Arc welding has not previously been used for such miniature devices because equipments have generally been insufficiently sophisticated to allow precise control of the arc to enable the low heat input levels necessary if device and instrument degradation is to be avoided. The new generation precision TIG equipment with solid state controls, however, enables the use of extremely low currents down to 0.1A with an accuracy of better than ±5% and with pulse shaping/control (1 to 500Hz, current high and low duration ratios of 10–90%). Consequently the welding of sheet stainless steel, nickel alloys and titanium of less than 0.1mm thickness is now feasible under production conditions.

Fig 10. Diaphram (0.075mm thick, 19mm diameter) welded by pulsed TIG using a solid state power supply

Similar advances have been made with resistance welding equipments. Until recently simple AC, DC (battery or capacitor discharge) power sources were employed. Solid state transistorized DC equipments are now available with adjustable up-slope and down-slope facility, with voltage, current, power and energy feedback control to facilitate the welding of difficult materials (eg heavily oxidised surfaces) and with current direction control for material combinations which show an excessive Peltier effect.

Nd-YAG laser welding and machining of precision components is being increasingly employed since, unlike EB welding, it can be performed in any appropriately selected atmosphere. Laser welding also causes less device distortion than resistance welding, suffers no problems of electrode wear, and offers greater freedom of design since weld size is not limited by the minimum practical electrode dimensions. The ability to use fibres for transmitting the laser beam also gives great flexibility. The fibres are very efficient at carrying the laser energy and losses from them are negligible, especially when compared to the 18% loss which occurs when the beam is fed in and out of the fibre. Thus fibres can be as long as is practical - 100m if necessary. This gives the major advantage that the laser itself can be at a distance from where the welding is occurring. This is especially useful when wanting to isolate the laser from hazardous areas or to use it in very clean areas (eg for medical applications and clean rooms). Also, the laser can be shared between many welding stations. One laser source can feed many fibres, allowing them to operate either simultaneously or, by using a distributor arrangement, sequentially. For example, spot welds can be made around a component by sending laser pulses down individual fibres positioned around the component, rather than by having to move the component beneath a welding head. This is particularly important for components which require critical alignment. Also, for parallel production lines performing spot welds, the laser pulse can be sent to the machine when the components are ready to be welded.

An example of the use of advanced Nd-YAG and resistance welding techniques is shown in Fig 11.

Fig 11. Opto-electronic device requiring advanced welding techniques for assembly. (Courtesy British Telecom Research Laboratories, UK)

The assembly operation for this opto-electronic device requires advanced welding processes including controlled DC resistance welding to minimise heat input and precise laser welding techniques to ensure consistent alignment of the optical fibre to the transmitter/receiver.

# PLASMA ARC WELDING AND CUTTING

K. NISHIGUCHI
Department of Welding and Production Engineering, Osaka University

Abstract
The present report was arranged to describe the process developments of plasma arc welding and cutting to consider their research direction. So, the contents are divided into three parts. At first, various kinds of the machine for the plasma arc welding and the plasma arc cutting which are used in the present time is classified by the arc mode and their structures and characteristics. In the second section, the developments of plasma arc welding, plasma cutting and their power source are described on the basis of the literature. These developments are respectively arranged into two chronological charts. In the last section, research history of the arc phenomena and process mechanism in plasma arc welding and cutting and current state of their studies are arranged to be described their future and research direction.
key word: Transferred arc, Semi-transferred arc, Non-transferred arc, Micro-plasma, Plasma arc, Plasma jet, Keyhole welding, Overlay welding.

## 1. Introduction

It has been 40 years since Burbon invented the plasma arc welding process in 1950, and also 35 years since the plasma arc cutting process was applied to cut aluminum alloys by Linde Co.(1955a).

The plasma arc welding had scarcely been paid attention before it was used for welding rocket cases in the middle of 1960s. So, the welding machine was not on the market until the end of 1960's. Studies on the plasma arc welding were progressed mainly in the field of aerospace industry after the adoption. Especially, USA took a leading position of the studies. In the USA, a plasma arc welding machine of variable polarity type has been developed in the recent year, and it adopted for welding the space shuttle made of aluminum alloys. Todays, pulsed plasma arc welding is used for all position welding of pipes and micro-plasma arc welding for the thin plates. The monitoring technique of the weldment for automated system and the know-how for high quality welding of various materials are an important problem in the near future.

On the other hand, the plasma cutting in the early years was used for a cutting method of stainless steels and aluminum alloys which are difficult to cut by the oxy-fuel gas cutting. After the end of 1960s, however, the application has been extended to the steel cutting field in shipbuilding factories. The cutting machine of the low current type is spread just in 1980s, although a low current type appeared in the latter half of 1960s. Since it was improved to be compact, light and cheap, it remarkably spread to the various fields. In Japan, 15 to 20 thousands units has been sold every year from the middle of 1980s. A type of plasma cutting machine with pure oxygen as a plasma gas has been also spread from this 1980s. In the field of plasma cutting, the studies on the improvement of the cut surface quality involving the bevel angle and the researches on the improvement of electrode lifetime are now required to be energetically performed.

Details of the developments, the current states, and future direction of the researches in plasma arc welding and cutting are described in the following sections;

## 2. The plasma process

### 2.1 Principle of operation

In the plasma arc welding and cutting, high temperature and high density plasma flow is usually used as a heat source, which is formed by a physical constriction of the arc plasma with a copper nozzle. Generally, the plasma arc is stroked in straight polarity where electrode is negative. Then, the arc voltage is pretty higher than that of GTAW of free arc state because the current path in arc is constricted into a relatively small diameter by the nozzle.

### 2.2 Torch operation modes

The plasma arc is divided into two operation modes. One is the non-transferred arc type as shown in Fig. 1 (a) which was introduced by Schönherr in 1906. This is often called plasma jet type. In this type, arc is produced between the constricting nozzle and the elec-

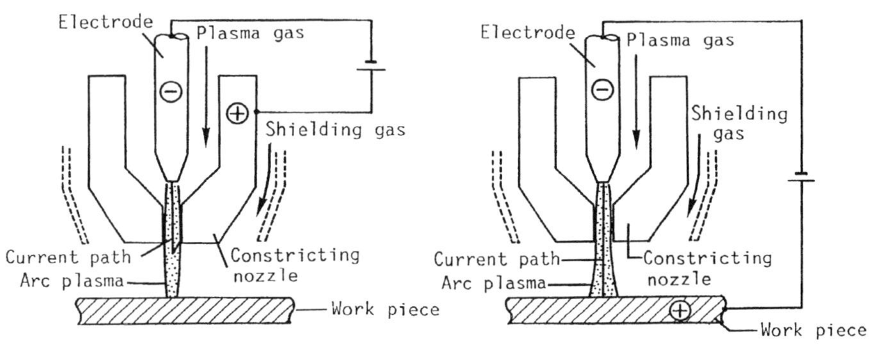


(a) Non-transferred arc. (b) Transferred arc.

Fig. 1 Basic modes of plasma arc operation.

trode inside of the nozzle, and the plasma gas is supplied in vortex flow. The plasma jet type, however, is not applied so well to the welding or the cutting. The jet type is chiefly used for the plasma spraying. Recently, this type begins to be merely applied to the cutting of plastic plate, too.

The other is the transferred arc type as shown in Fig. 1 (b) and called plasma arc. The plasma arc is generated between an electrode within nozzle and a workpiece by the high temperature and conductible plasma flow caused by a pilot arc generated between the electrode and the nozzle. The plasma gas is supplied in vortex or axial flow. This type makes the energy density at heat source higher, compared with the plasma jet type. Moreover, in this type, anode or cathode spot energy is available for melting of the workpiece. This is the reason why the plasma arc type is exclusively used to the welding and the cutting.

The transferred arc, however, becomes unstable if the arc current remarkably reduces in about 10 A or below since the arc voltage in such a low current condition decreases with the arc current (1971a). In this case, a few currents between the nozzle and the electrode are required to keep the transferred arc stable. This type is called micro-plasma or semi-transferred arc.

Table 1 shows the arrangement of the relationship between the three kinds of arc operation modes and the various plasma arc welding and cutting methods.

Table 1 Classification of the plasma process by arc modes.

Plasma process
- Non-transferred arc ( Plasma jet )
  - Plasma jet cutting (PJC)
  - Micro-plasma jet cutting (Micro-PJC)
- Semi-transferred arc (Micro plasma arc)
  - Micro-plasma arc welding (Micro-PAW)
  - Micro-plasma arc cutting (Micro-PAC)
- Transferred arc (Plasma arc)
  - Soft plasma arc for overlay welding
  - Plasma arc keyhole welding (PAW)
  - Plasma arc cutting (PAC)

## 2.3 Operating conditions

The apparatus of plasma arc welding is quite similar to that of the plasma arc cutting, although these methods are distinct from each other. In the recent years, there has been a plasma cutting torch even with the shielding gas called "assist gas" (1988b). The difference between two processes exists only in the kinds of plasma gas and the gas flow rate during the operation, resulting in the arc voltage. The load voltage in the cutting operation is three or five times as high as that in the welding one.

In the plasma arc cutting, $Ar+H_2$, $N_2$, air, $O_2$ etc are selected for plasma gas and the gas flow rate is set in high momentum conditions to actualize the dross-free cutting. On the other hand, in the plasma arc welding, Ar, $Ar+H_2$ or Ar+He is adopted as the plasma

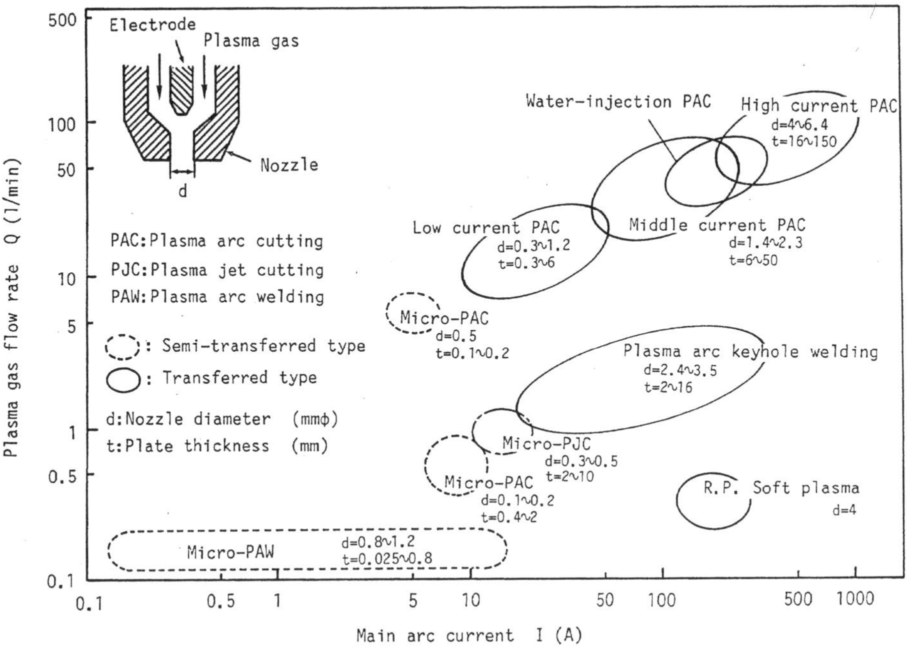


Fig. 2 An arrangement of operating conditions in various plasma processes and their applicable range in plate thickness.

gas. And Ar, $Ar+H_2$, $Ar+CO_2$, or Ar+He is the shielding gas.

A power source of constant current type is recently in both of the methods.

Maximum usable current value in the plasma arc operation must be below the critical series arc current, since the maximum current is limited by the double arc phenomena. The critical current is governed by the nozzle diameter, nozzle constriction length and gas flow rate. Usually, large diameter nozzle is recommended for high current condition due to thick plate welding and cutting.

Figure 2 shows common operating condition range of the each plasma process classified in Table 1. The figure is made from a lot of catalog and manual data. The gas flow rate and the arc current was selected as the axes. The adoptable plate thickness of workpiece and the favorable nozzle diameter are displayed in each plasma process, too.

## 3. Recent developments

### 3.1 Plasma arc welding process

In early years, the plasma arc welding was chiefly developed in the keyhole welding method of electrode negative polarity, as shown in Fig. 3. The micro-plasma arc welding with continuous pilot arc to make the transferred arc stable has been progressed after the end of

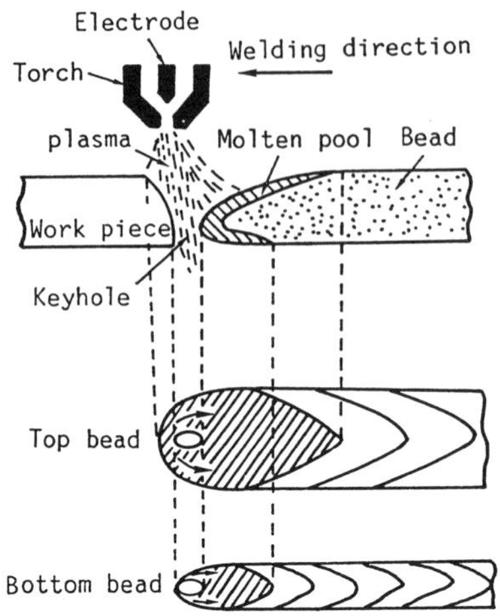

Fig. 4 A cross section of overlaid metal with a reverse polarity soft plasma arc.

Fig. 3 Schematic illustration of plasma arc keyhole welding.

1960s. Remarkable advance of the plasma arc welding process was recognized in 1980s. In the recent ten years, the pulsed plasma arc welding with current pulsation of 2 to 10 Hz has been applied for all position welding of pipes (1974c), and the variable polarity plasma arc welding of square AC wave has been developed to weld aluminum alloy used in space shuttle external tank (1984).

A soft plasma arc overlay welding with a large diameter nozzle was also developed in the latter half of 1970s. In the method, the distribution of cathode spot on the base metal is controlled only by

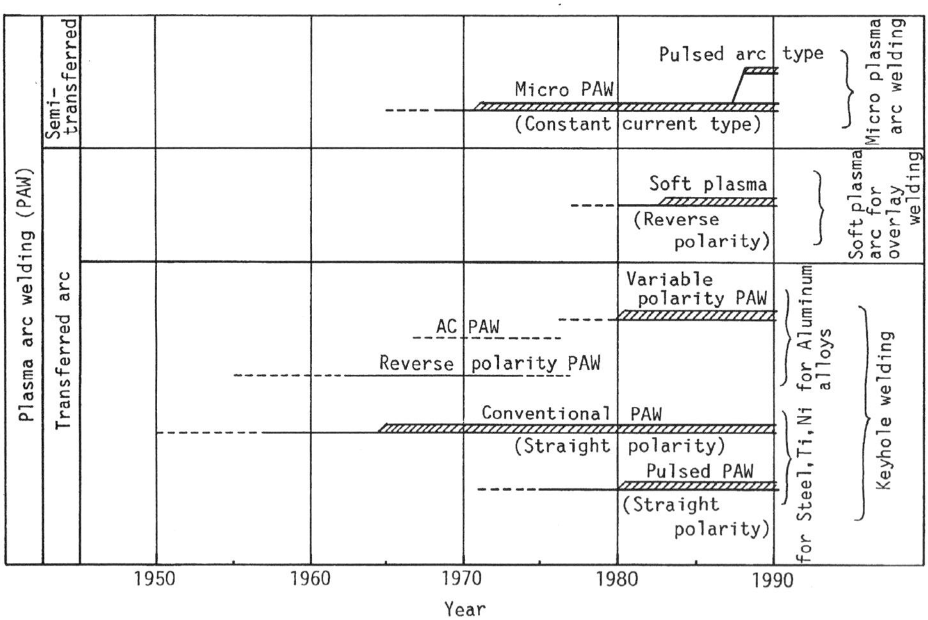


Fig. 5 History of the developments in the plasma arc welding.

the composition of shielding gas. As seen in Fig. 4 (1977), the overlay welding is now actualized with the dilution < 0.1%.

Figure 5 shows the history about the development in various kinds of plasma arc welding. The period of researching stage is expressed with the broken lines. The solid line expresses practical use stage. The hatched zones express the period in which the process seems to be widely utilized.

### 3.2 Plasma cutting process.

The plasma arc cutting process has already been spread by the first half of 1960s, and in early years it has been utilized only as the cutting method of stainless steel and aluminum alloy of relatively thick plate in high current condition of 500 to 1000 A with $N_2$ and $Ar+H_2$ for plasma gas. After that, the former $N_2$ plasma cutting process developed to water-injection type for improvement of environment for worker besides the cut quality.

The plasma arc cutting of steel with air for the plasma gas and Zr-inserted copper rod for electrode was spread after the first of 1970s, although the study had been started in the early years(1962). The air plasma cutting machines mounted on NC apparatus were introduced to many shipyards with middle current torch of 150 - 250 A. Recently, the oxygen plasma arc cutting machine becomes to be used for steel to avoid the nitrogen occlusion of the cut surface(1974b).

The low current type below 50 - 100 A was rapidly spread to various fields, since some light and cheap plasma cutting machines using low current were sold in the middle of 1980s. In particular, the plasma cutting machine has been popularized after the contact cutting type below 50 A was on the market for manual operation.

Figure 6 shows the recent situation about the possession of the plasma cutting machines in Japan. This is based on an investigation with a questionnaire performed in 1989 by the Japan Welding Engi-

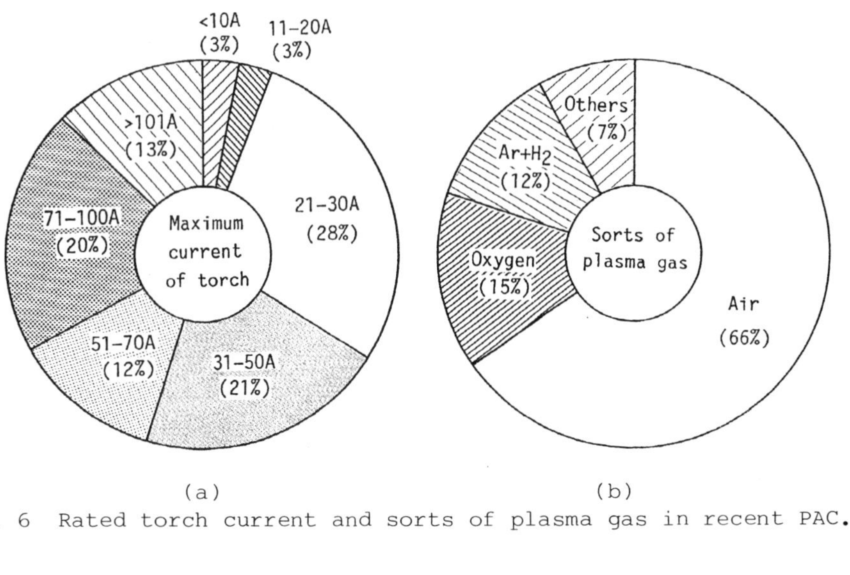


Fig. 6 Rated torch current and sorts of plasma gas in recent PAC.

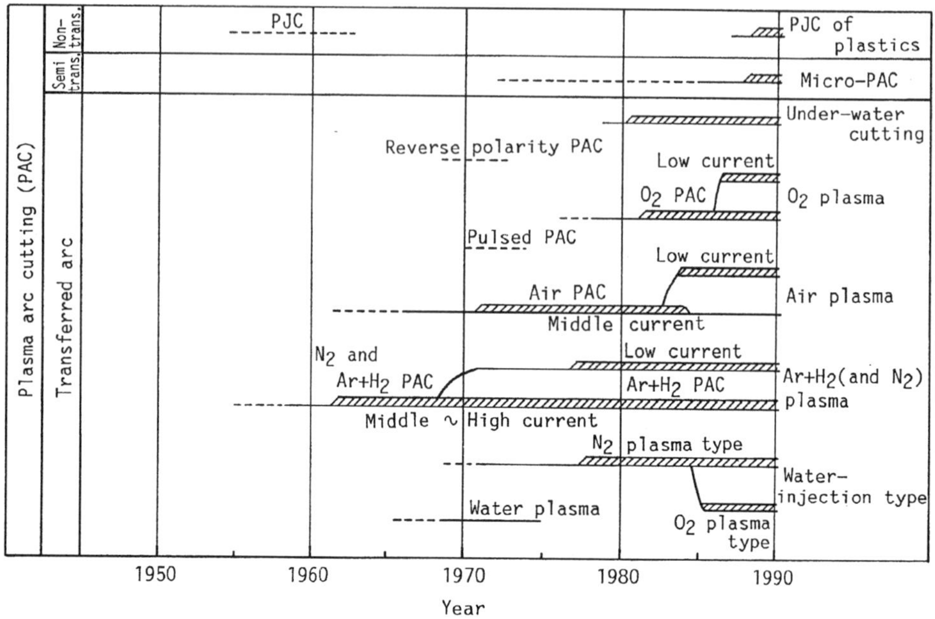


Fig. 7 History of the developments in the plasma arc cutting.

neering Society. The low current below 50 A are close to 55 %, and the oxidizing gas is adopted for the machines of about 80 %.

In addition, micro-plasma arc cutting machines for thin plate and the low current plasma jet cutting machines for plastics began to be utilized in recently (1988b). The history of the development is shown in Fig. 7 with the same manner as in Fig. 5.

### 3.3 Electric Power Source

Recently, power sources used to the plasma arc welding and cutting machines have rapidly been evolving. In the early years, the satu-

Table 2 Construction of power sources.

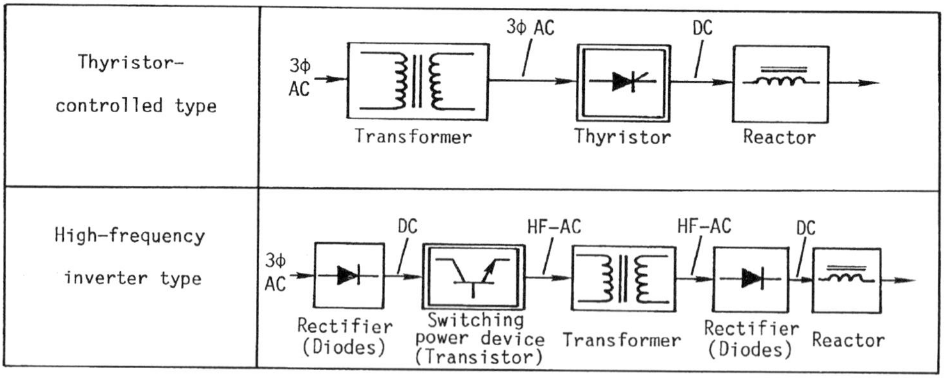

rating characteristics of reactor was utilized to control the arc current. And, the power source changed to the thyristor controlling type in 1970s. The inverter type appears in the middle of 1980s. The compositions of the thyristor-controlled type and the inverter-controlled one are shown in Table 2.

The inverter type was applied to the low current air plasma cutting machines below 50 A and to the pulsed plasma arc welding ones since the newest power source is light and easily controllable, although the type is a little expensive. The thysistor type is adopted in the rest field. The transistor chopper type is scarcely used now.

The power source used in the variable polarity plasma arc welding was first made of the thyristors but is now changed to the double inverter type setup in both of primary and secondary sides.

## 4. Current State of Research and Research Direction

### 4.1 Plasma Arc Phenomena

It is well known that the theoretical studies on the "plasma" began in a report described by Langmuir in 1923. However, studies on the transferred arc directly concerning to the plasma arc welding and cutting were started from Maecker's study on the wall-stabilized arc in 1955 and Gage's research on the gas-stabilized arc in 1959. In 1955, Maecker explained an accurate mechanism of magnetic pumping. At the same time, he described the double arcing phenomena by that the maximum current is limited in the plasma arc, too.

Fundamental studies to examine the temperature, the flow velocity and the arc efficiency etc of plasma arc were energetically performed from 1960s to the first half of 1970s. After that, these researches gradually declined. However, we cannot say that the governing factors of the plasma arc behavior are identified. And we have not easily control the plasma arc yet.

In the author's laboratory, a numerical simulation of plasma arc behavior and an algorithm with a reverse problem technique for the information obtained from the mutual interference between the plasma arc and the melting front in the workpiece are now carried out for explicating the governing factor of the plasma arc behavior.

### 4.2 Plasma arc welding phenomena

The study on the plasma arc welding was started from the keyhole welding phenomena. From the latter half of 1960s to the beginning of 1970s, the sustaining mechanism of molten pool and the weldable region were examined in the plasma arc keyhole welding. The maximum sustainable bead width is shown in Fig. 8 with experiments (1971a). The critical width in the plasma arc keyhole welding of thin plate is different from that in usual GTAW with no keyhole action, since the maximum bottom bead width in the keyhole welding determined only in thin plate condition becomes small by that the molten metal existing on both wall molten metal in welding crater becomes difficult to be bridged with each other in such a thin plate condition.

The explication of the bead formation phenomena in the plasma arc keyhole welding has been developed by the beginning of 1970s, taking account of the effect of keyhole action and the secondary melting action by the molten metal. Then, a fundamental model of the bead formation has been established in straight polarity where the electrode is negative. The cathode spot behavior on the workpiece in reverse polarity was researched with a water-cooled copper electrode besides the large size tungsten electrode in the beginning of 1970s.

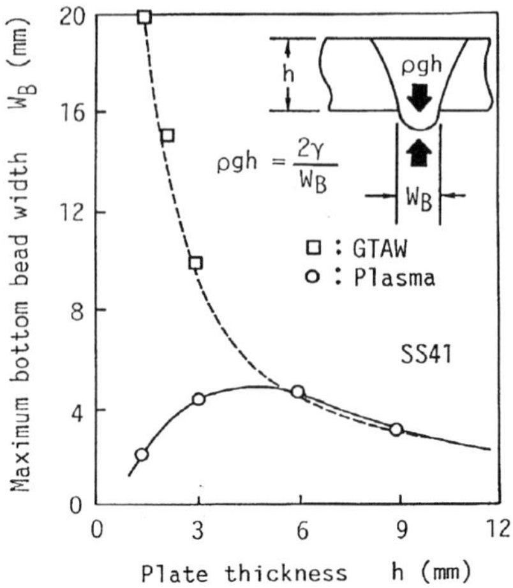


Fig. 8 Relations between maximum bottom bead width and plate thickness.

Many fundamental studies on the micro-plasma arc welding of semi-transferred type were also performed in the beginning of 1970s, too. The studies on pulsed plasma arc welding and variable polarity plasma arc welding were advanced after 1970s. Todays, it was clarified that a sequence of one-pulse/one-keyhole is suitable for the all position welding of pipes (1986a), and that a condition of a cycle of 20 ms of electrode negative polarity, followed by 3 ms of electrode positive polarity is favorable to use the variable polarity plasma arc welding of aluminum alloys (1984a).

In the future, several problems will be important; the accumulation of know-how in the plasma arc welding for various materials, the systematization of the welding process, the improvement of the welding stability and the monitoring technique for the automating the welding process.

### 4.3 Plasma arc cutting phenomena

The studies on the plasma arc cutting phenomena were started from the choice problem of plasma gas. The concept as to the governing factors against the cutting ability and dross free condition had been arranged by 1970s. The kerf formation mechanism was examined, too. However, the problems as to the bevel angle and the top edge roundness have remained until now.

Todays, a part of these problems can quite be suppressed by using a newest oxygen plasma arc cutting machine. High quality cuts, which is called fine cut, as good as that in the laser cutting can be obtained with the new machine as shown in Fig. 9 (1989a). The idea of high quality cutting will be extend to the thick plate, although the now adoptable range of plate thickness is in thin plate below about 6 mmt.

Since the low current type with oxidizing gas was rapidly spread, the studies on the electrode lifetime has been performed. An ar-

(a) Kerf shape
(h=2.3mm, I=12A, v=1.1m/min)

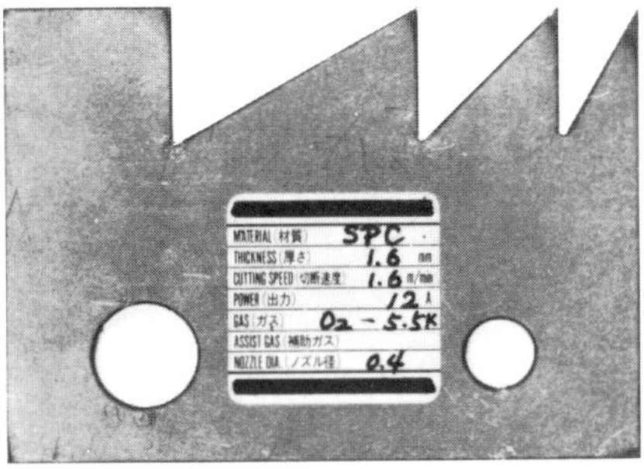

(b) Cut shape
(h=1.6mm, I=12A, v=1.6m/min)

Fig. 9 High quality cuts by a newest oxygen plAsma arc cutting.

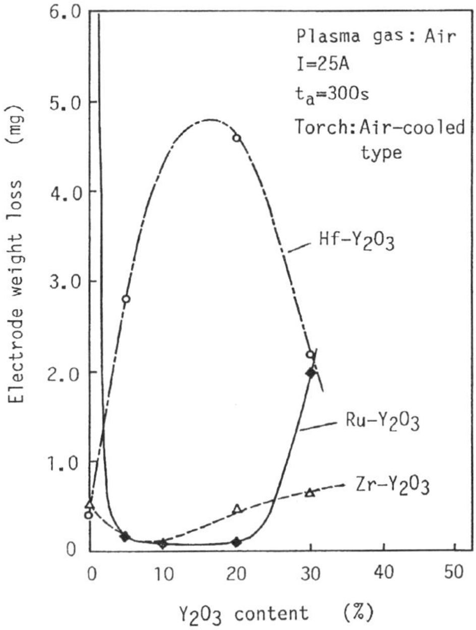


Fig. 10 Effects of using durable electrode on improving electrode lifetime.

ranged result is shown in Fig. 10 (1989b). The figure indicates the relationship between the composition of durable electrode materials inserted into copper rod and the electrode lifetime under a low current condition, and suggests that the new material is favorable to extent the lifetime. In the future, the study will be necessary to apply the large current conditions.

## References

Schönherr, O. (1909) Electrotechnische Zeitschrift, 22, 365.

Langmuir, I. and Tonks, L. (1929) Phsy. Re., 33, 195.

Linde Co. (1955a) Linde universal inert-gas tungsten-arc cutting of aluminum. Welding J., 34, 1097-1098.

Maecker, H. (1955b) Plasma streams in electric arcs as a result of self-magnetic compression. Z. Physik, 141.

Gage, R. M. (1959) The principle of the modern arc torch. Welding J., 38, 959-962.

Browing, J. A. (1962) Thermal-air cutting process and economics. Welding J., 41, 132-140.

Browing, J. A. and Hazard, C. (1963) Quality determinants in plasma cutting process. British Welding J., 462-469.

Filipski, S. P. (1964a) Plasma arc welding. Welding J., 43, 937-943.

O'Brien, R. L. et al. (1964b) Advance in plasma arc cutting. Welding J., 43, 1015-1021.

Shamblin, J. E. and Amstead, B. H. (1964c) Plasma arc cutting, Welding J., 43, 470s-472s.
Cooper, J. P. and Browning, J. A. (1965) Recent developments in plasma welding. Welding J., 44, 268-276.
Miller, H.R. and Filipski, S.P.(1966a) Automated plasma arc welding for aerospace and cryogenic fabrications. Welding J., 35,493-499.
Levin, H. L. (1966b) Plasma arc cutting - A review. Welding & Metal Fab., 34, 124-130.
Gagen, A. F. and Daiker, A. L. (1966c) Arc behaviour in the plasma cutting of metal. Svar. Proiz., No.12, pp11-12.
Norenko, V. P. and Spicheko, N. I. (1966d) A plasma arc cutting torch with the arc stabilized with air and water. Svar. Proiz., No.2, pp29-31.
Skinner, G. M. and Wickham, R.J. (1967) High quality plasma arc cutting and piecing. Welding J., 46, 657-664.
Dudoko, D. A. and Kornienko, A. N. (1968a) The thermal efficiency of the fusion process in plasma arc welding. Avt. Svarka,No.7,18-20.
Wilson, E. M. (1968b) Micro-plasma arc welding of thin high strength steel strip. Welding & Metal Fab., 36, 133-139.
Sivirkin, V. F., Martynov, V. I. and Rogachef, N. M. (1968c) Experiments on the shunting mechanism of the anode spot of an arc with an axial flow of gas. Svar. Proiz., No.10, pp12-13.
Kudinof, V. V. et al. (1970a) Heat exchange between a plasma arc and the metal to be cut. Svar. Proiz., No.2, pp2-3.
Malin, V. E. (1970b) A study of the thermal characteristics of the constricted A.C. arc. Svar. Proiz., No.9, 3-5.
Nishiguchi, K. (1971a) Plasma arc welding. J. of the Japan Welding Society, 40, 1206-1219. (in Japanese)
Couch Jr, R. W. and Dean Jr, D. C. (1971b) High quality water-arc cutting. Welding J., 50, 233-236.
Smart, M. D. and Pinfold, B. E. (1971c) Comparison of modulated and unmodulated current plasma-arc welding. Welding & Metal Fab., 39,332-341
Bunting, K. A. (1972) Metal-foil cutting by plasma, Metal Const. & British Welding J., 88-92.
Nishiguchi, K and Matsuyama, K (1974a) Phenomena of high speed cutting. J. of the Japan Welding Society, 43, 805-817. (in Japanese)
Slivinskii et al. (1974b) Effect of the parameters of air-plasma cutting conditions on the weld edges with nitrogen. Avt. Svarka, No.10,58-59.
Lundin, C. D. and Ruprecht, W. J. (1974c) Pulsed current plasma arc welding, Welding J., 53, 11-19.
Tomsic, M. J. and Jackson, C. E. (1974d) Energy distribution in keyhole mode plasma arc welds. Welding J., 109s-115s.
Shimada, W. et al. (1977) High quality overlay welding with reverce polarity soft plasma arc. Com. of Welding Arc Physics of J.W.S., 77-360. (in Japanese)
Nishiguchi, K. and Matsuyama, K. (1979) Kerf formation and dross adhesion in plasma arc cutting. Welding in the World, 17, 232.
Shimada, W., Hoshinouchi, S., Hiramoto, S. and Ukai, J. (1980) Study on characteristics of reverse polarity soft plasma arc and its apprication to overlay welding. IIW, Doc. No. 212-480-80.

Matsunawa, A (1983) Role of surface tension in fusion welding (Part 2). *Transaction of JWRI*, 12, No.1, 123-130.
Tomsic, M. and Barhost, S. (1984a) Keyhole plasma arc welding of aluminum with variable polarity power. *Welding J.*, 63, 25-32.
Nunes, A. C. Jr, Bayless, E. O. Jr, Jones, C. S. III, Munafo, P. M., Biddle, A. P. and Wilson W. A. (1984b) Variable polarity plasma arc welding on the space shuttle external tank. *Welding J.*, 63, 27-35.
Maruo, H. and Hirata, Y. (1985a) Study on pulsed plasma arc welding phenomena. *Com. of Welding Arc Physics of J. W. S.*, 85-579. (in Japanese)
Brosilow, R. (1985b) Advances in PAC speed plate cutting. *Welding Design & Fablication*, No.1, 30-32.
Haneda, M. et al. (1986a) Study on all-position plasma keyhole welding. Quarity *J. of the Japan Welding S.*, 4, 300-305. (in Japanese)
Yamamoto, H. (1986b) Recent trends of low current air plasma cutting. *J. of the Japan Welding Society*, 55, 361-369.
Cuny, F. (1986c) Plasma arcs weld carbon steel parts. *Welding design & Fablication*, No.2, 50-52.
Schultz, J. F. (1986) TIG process with dual shield: intermediate process between TIG and plasma arc welding. *Welding in the World*, 24, 24-258.
Kyseica, S. (1987a) High-frequency Reverting arc switch for plasma arc welding of alminum. *Welding J.*, 66, 31-35.
Matsuyama, K. (1987b) Electrode wear phenomena in oxygen plasma. *Plasma division in Japan Welding Eng. S.*,G4-87-262.(in Japanese)
Frappier, M. B. (1988a) Plasma arc cutting power supplies explained. *Welding J.*, 67, 48-50.
Hasegawa, (1988b) Neoplasma arc & jet. *Plasma division in Japan Welding Eng. S.*, G4-88-275. (in Japanese)
Sakakibara, J. et al. (1988c) Heat efficiency in plasma jet cutting. *Preprint of the national meeting of J.W.S.*,42,272-273.(Japanese)
Craig, E. (1988d) The plasma arc process. *Welding J.*, 67, 19-25.
Bashenko, V. V. and Sosnin, N. A. (1988e) Optimization of the plasma arc welding process. *Welding J.*, 67, 233s-237s.
Inoue, M. (1989a) Komatsu fine plasma. *Plasma division in Japan Welding Eng. S.*, G4-89-289. (in Japanese)
Matsuda, H., Ushio, M. and Kusumoto, K. (1989b) Study of new cathode material for air-plasma. *New Tec. Division in Japan Welding Eng. S.*, CT3-10-89. (in Japanese)
Richardson, I. M. (1989c) Plasma welding - the current status. *Welding & Metal Fab.*, 57, 185-190.
McGough, M. S., Aurtin, W. E. and Knetl, G. J. (1989d) Underwater plasma arc cutting in three mile island's reactor. *Welding J.*, 68, 22-27.
Woodard, L. D. (1989e) Microplasma arc welds switches. *Welding Design & Fablication*, No.3, 67-69.
Madsen, H. and Hausler, H. (1989f) Plasmaschneiden unter Wasser mit den Plasmaanlagen PA 12 UWP und PA 50 UWP. *Schweisstechnik* (Berlin), 39, 52-56.

# THE ROLE OF COMPUTERS IN JOINING PROCESSES

K.E. EASTERLING
Department of Engineering Materials, University of Luleå

Abstract
This paper is mainly concerned with the use of computers for predicting microstructural and property changes that occur in welding. Welding is a complex metallurgical process involving a severe thermal cycle, rapid solidification of the melt, precipitate dissolution, grain growth and phase transformation in the heat affected zone of the base metal. Computing is essential in the development of models which attempt to provide reasonable mathematical descriptions of all of these phenomena such that useful output data is rapidly obtained. Such programs should also ideally be interactive with the operator, allowing changes in input data as necessary. Examples are given of the application of computer programs to microstructure and property prediction in welds and welding tests.

Keywords: Computer, Software, Welding, Joining, Modelling, Microstructure, Properties, Prediction.

## 1. Introduction

The role of computers in welding was discussed extensively at the 1988 IIW meeting in Vienna (1988), and indeed, it was also the subject of the Hondremont lecture of the same meeting (Feder, 1988). As can be seen from the proceedings of the 1988 meeting, four main areas of computer-aided welding were identified, these being:

1) Design, calculation and prediction models for metallurgical processes.

2) Inspection and in-service monitoring.

3) Fabrication, quality assurance.

4) Expert systems, data banks and future possibilities.

This paper is mainly concerned with the theme addressed in (1), i.e. calculation and prediction models for joining processes.

For many years prediction models of weldability were based largely on empirical formulae derived by curve-fitting to experimental results (see e.g. Hrivnak, 1978). A good reason for this approach was that it at least afforded welding engineers a simple but effective measure of potential cracking problems under well defined conditions of material composition and welding process. Thus, while metallurgists have built up a good understanding of the various physical phenomena occurring in conjunction with welding, equations describing these phenomena were, until fairly recently sufficiently complex as to discourage their regular (and reliable) adaptation by welding engineers. In the past 5-6 years or so, however, this situation has changed somewhat, partly by the development of plausible prediction models, and not least by the increased availability and capacity of good microprocessors (Fig 1). Indeed, progress in the field has been such that at the 1988 IIW meeting, a subgroup of Commission IX (b) was initiated with the purpose of encouraging and collecting work on the analysis and modelling of welding processes. As a forum for this approach, table I illustrates some of the possible themes to be addressed.

For any given set of input parameters arising from the welding process itself, type of material being welded, the part of the join of interest (weld metal or HAZ) etc., equations can be sought or developed describing the various individual phenomena to be treated, e.g. shape of the weld thermal cycle (or cycles), types of precipitate or inclusions involved, role of hydrogen, effect of microstructural changes or properties, etc. It is in assembling all of these various input equations together, allowing them as it were to interact with oneanother to produce useful data as output, that the use of a computer is invaluable.

Table 1. Input and output data for modelling welding processes and associated phenomena

| Applications (input parameters) | Phenomena (equations sought) | Objective (output data) |
|---|---|---|
| Weld process<br>-arc<br>-laser<br>-electron beam<br>-resistance<br>-pulsed welding<br>-brazing<br><br>Base materials<br>-steels and cast iron<br>-stainless steels<br>-Al-alloys<br>-Ti-alloys, etc<br>-metal/ceramic bonds<br><br>Weld metal<br>-solidification structure<br>-phase transformation<br>-inclusions' role<br>-precipitation<br><br>HAZ<br>-microstructures<br>-properties<br><br>Reheat structures<br>-multi-run welds<br>-stress relieving | Arc physics<br><br>Heat flow of weld cycle<br>-continuous<br>-pulsed<br><br>Transformations<br>-kinetics<br>-morphology<br>-crystallography<br>-role of inclusions and precipitates<br><br>Precipitation<br>-dissolution<br>-coarsening<br>-re-precipitation<br><br>Grain growth<br>-recrystallization<br><br>Hydrogen<br>-diffusion<br>-distribution<br>-role of defects<br><br>Reheat effects (multi-run welds)<br>-grain-refining<br>-tempering<br><br>Properties<br>-hardness<br>-strength<br>-toughness<br>-residual stresses | Predictive equations and diagrams<br><br>Microstructure prediction<br><br>CCT behaviour<br><br>Property prediction<br>-hardness<br>-toughness<br><br>Duplex microstructure prediction (stainless steels)<br><br>Cracking susceptibility<br>-cold cracking<br>-solidification cracking<br>-reheat cracking<br>-liquid cracking<br>-lamellar tearing<br><br>Development of intelligent systems<br><br>Documents (IIW)<br><br>Software development for general use<br><br>Technique development |

The final objectives, listed in the right hand column of Table 1 represent the computer output data. These may be in the form of CCT diagrams, hardness changes, cracking susceptibility warnings, or simply microstructural predictions, e.g. grain growth in the HAZ, ratio of ferrite to austenite in stainless steels, etc. Obviously, the ultimate objective is to produce computer software that can help the welding engineers optimise the joining process by e.g. using combinations of process and material/design parameters that best avoid welding problems, cracking or high residual stress. Ultimately, programs of this type may even be in the form of "intelligent systems", giving on-line feedback and allowing adjustments to the process as it occurs.

Some useful developments in these general directions have been made; It is the purpose of the present paper to illustrate with a few examples, the sort of progress that is actually going on.

## 2. Selected examples of computer-assisted modelling

### 2.1. Weld metal CCT diagrams

This is a difficult assignment, in that the thermal cycle of the melt zone of a fusion weld is hard to characterise mathematically (see, however, the papers of Eager (1990), Goldak and Bibby (1990). Physically, the melt pool is complex because of the huge temperature gradients involved, unusual chemical reactions that occur at peak temperatures, the complex fluxes and inclusions in the melt, melt turbulance and of course certain fundamental differences between the various welding processes. Thus modelling of weld metal solidification structure is not a problem for the faint of heart! Bhadeshia and co-workers, 1985, 1989, have however, devoted their considerable energies at overcoming some of these problems, and even succeded in developing (albeit with computer assistance) a useful phenomenological model, the scheme of which is illustrated in Fig 2. Wisely avoiding the high temperature gradient problems below the arc, this model concentrates on the cooling cycle at the fusion line, making certain assumptions concerning initial and dendritic/austenite grain size. This approach thus allows use of an analytical thermal cycle equation, which, taken together with regular eutectoidal transformation kinetics, enables prediction of phase distribution. A comparison between theoretical prediction and experimental measurement is shown in Fig 3, giving good agreement.

### 2.2. HAZ microstructure and properties

From a computing point of view, this can be broken up into the following stages, using a microalloyed steel as the base metal (see e.g. Ion et al 1984):

Fig 1. Bridging the gap between computer simulation and real welding.

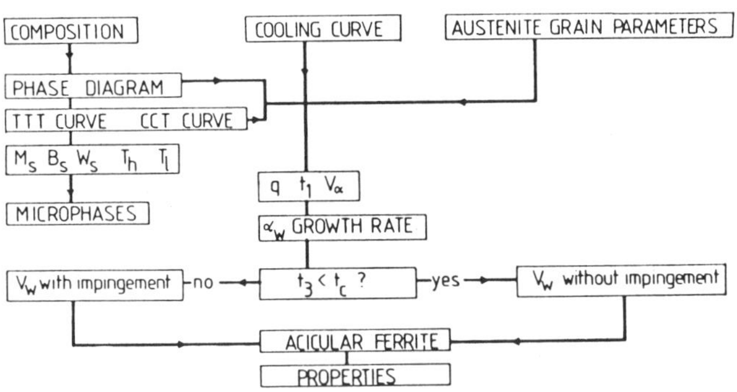


Fig 2. Schematic of computer program for calculating eutectoidal products in steel weld metal, after Bhadeshia et al (1985).

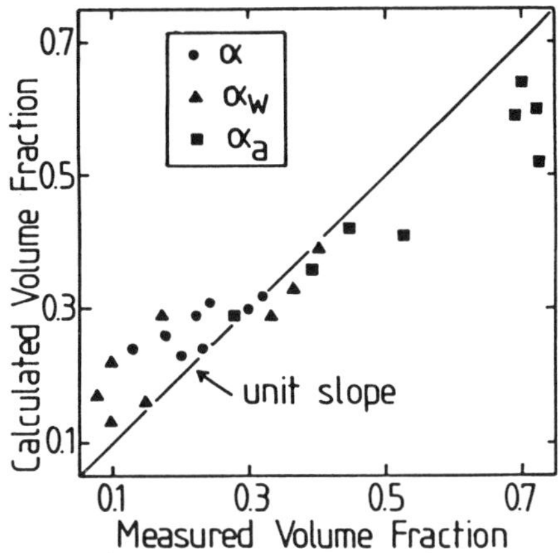


Fig 3. Comparison of calculated and measured values of various ferrite products in steel weld metal, after Bhadeshia et al (1989).

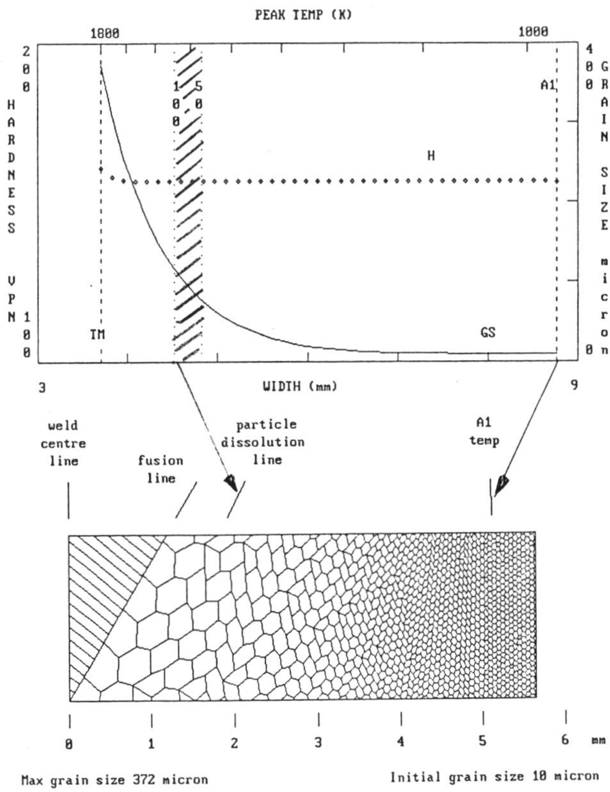


Fig 4(a-b). Computer readouts of austenite grain growth and final hardness in the HAZ of a Nb-microalloyed steel.

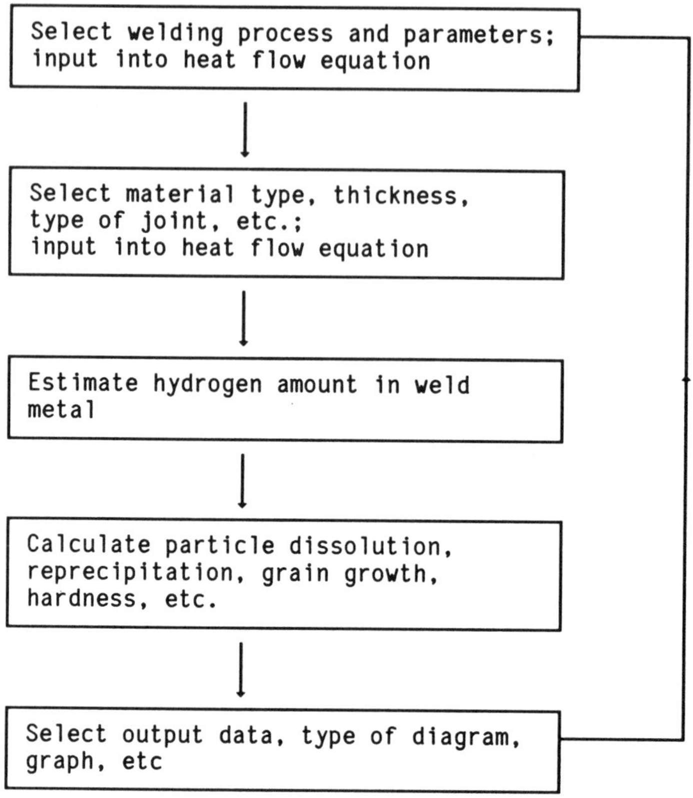


Examples of the type of output data produced by computer programs are illustrated in Fig 4 a and b. Fig 4(a) shows changes in prior-austenitic grain size and hardness as a function of distance from the fusion line. The shaded zone is the calculated temperature band of carbo-nitride dissolution, temperatures above which considerable austenite grain growth occurs. Other output giving hydrogen distribution in the HAZ, as well as cold crack susceptibility are given in this program. A computed sketch of grain size distribution in the HAZ is illustrated in Fig 4(b); grain sizes are not exactly to scale, although actual grain sizes are given. Computed "micrographs" like this, however, do have the advantage of presenting welding engineers with a reasonable physical picture of the weld zone, and they possess a certain pedagogical value, too. The program as a whole is written in such a way that it is easy for the welding engineer to "interact" with the microprocessor, such as to change input parameters at will, thus optimising the final weld.

2.3. Modelling welding tests
The concept of applying modelling to standard welding tests is attractive in that it may help refine and optimise the test, hence making it more reliable. One example has been given by Ion and Easterling 1985 for modelling implant tests. As illustrated in Fig 5, this test comprises of a test bar, placed in a base plate and then welded from above such that a notch in the bar ideally located in the grain growth zone (when loaded) acts as a stress raiser. The test is widely used as a means of testing for hydrogen induced cold cracking in the HAZ of welded samples subjected to tensile stresses. The object of the computer-aided modelling in this case is to help optimise the location of the notch, this being a sensitive parameter in hydrogen cracking. Ideally it should be located exactly at the grain growth zone of the HAZ. The program developed to do this was thus a development (sub-routine) of that discribed in 2.2. (above), and this is shown in Fig 6. Examples of the resulting computer outputs are illustrated on the computer console in Fig 1. A readout showing the ideal notch location in a welding diagram of variable input energy is shown in Fig 7. Another example of this application of computer prediction to cracking susceptibility for the case of the Tekken test has been given by Andersson (1981).

2.4. Residual stresses in welds
This has been the subject of intense effort over many years. Continued improvements in computer capacity, coupled with considerable ambitition and enthusiasm on the part of the modellers, have recently produced impressive examples of residual stress calculations applied to quite realistic applications. An elegant example is that given by Karlsson and co-workers, 1989. In this case, deformation strains and stresses during the butt welding of a pipe-line are calculated using FEM. The geometry of the pipe, together with the associated FE mesh and calculated hoop stresses are illustrated in Fig 8 (a-b).

## 3. Concluding remarks

Recent examples given in this paper of computer modelling, as applied to welded joints, illustrate the growing use, and usefullness, of this branch of welding research. Obviously, computer models are only as good as the data that is put into them; they can hardly substitute for actually carrying out a real weld, and are only a back-up substitute to a well designed real-weld test. However, computer software is beginning to appear, even at shop-floor level, as a means of helping engineers optimise weld properties. It is also an excellent way of making engineers and designers aware of the complexity of welding, including the many input parameters involved. Likewise, feedback from design engineers and welders to the modellers is equally important in helping them appreciate the special problems of real welding.

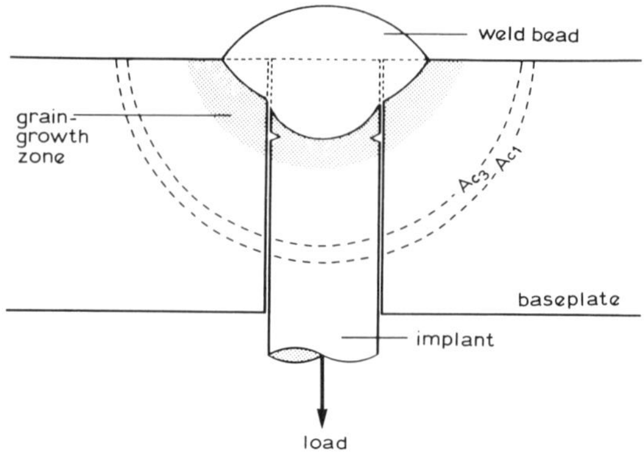


Fig 5. Test set-up for implant testing.

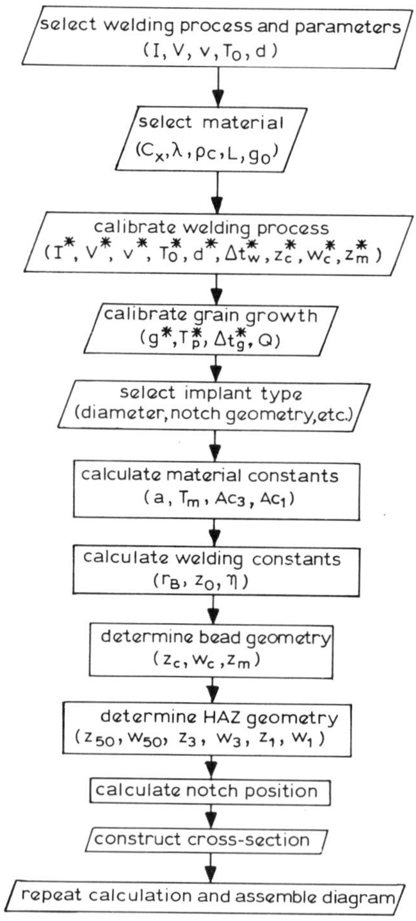


Fig 6. Schematic of computer program for predicting notch position in implant test, after Ion and Easterling, 1985.

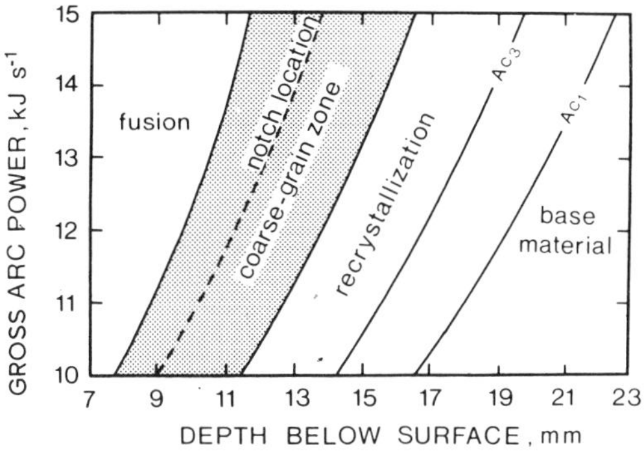


Fig 7. Computer plot of notch position as a function of weld input energy and depth below surface of implant test place, after Ion and Easterling, 1985.

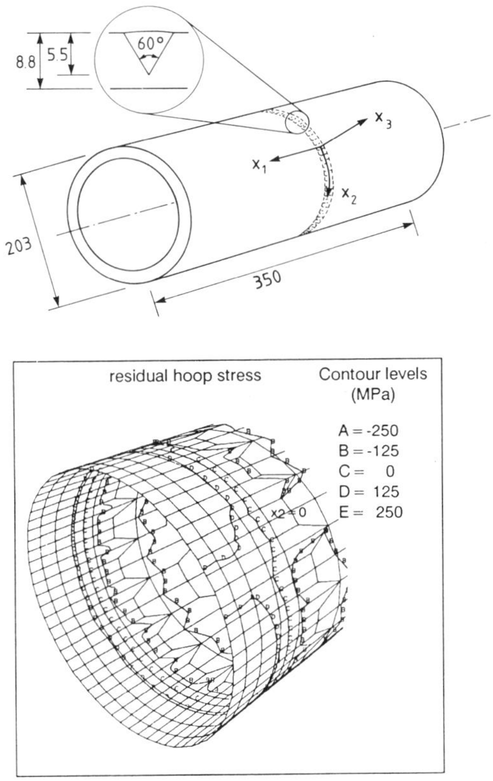


Fig 8(a-b). Test set-up for a butt-welded pipeline (a), showing FE mesh and computed results (b); after Karlsson et al (1989).

References

Andersson, B. (1981), Hydrogen cracking in weldments, Ph.D. thesis, Chalmers Univ. of Technology,

Bhadeshia, H., Svensson, L-E, Gretoft, B, (1985), A model for the development of microstructure in low-alloy steel (Fe-Mn-Si-C) weld deposit, Acta Met., 33, 1271-1283.

Bhadeshia, H., Svensson, L-E., Gretoft, B., (1989), Prediction of microstructure of the fusion zone of multi-component steel weld deposits, Proc. ASM Conf. Trends in Welding Research, Gatlinburg, Tennessee (Ed. S.A. David), p. 225-229.

Eagar, T., (1990) Physics of the arc, these proceedings (in press).

Feder, D.K., (1988), Computers in welding technology - a look at applications and potentials, IIW Conf. Hondremont Lecture, (Pergamon Press), p xvii-xlviii.

Goldak, J., Bibby, M., (1990), Heat flow and fluid flow considerations, these proceedings (in press).

Hrivnak, I., (1978), The mutual relationship and interdependence of developments in steel metallurgy and welding technology, welding in the world, 16, 1.

IIW Proceedings (1988), Weld quality: the role of computers, Pergamon Press.

Ion, J.C., Easterling, K.E., Ashby, M.F., (1984) A second report on diagrams of microstructure and hardness for heat-affected zones in welds, Acta Met., 32, 1949-1962.

Ion, J.C., Easterling, K.E., (1985), Computer modelling of weld-implant testing, Mat. Sc. and Tech., 1, 405-411.

Karlsson, L., Jonsson, M., Lindgren, L.E., Nasstrom, M., Troive, L., (1989), Residual stresses and deformations in a welded thin-walled pipe, Am. Soc. Mech. Engr., 173, 7-11.

PART TWO

# Arc Physics, Equipment and Controls

# THE PHYSICS OF ARC WELDING PROCESSES

**T.W. EAGAR**
Department of Materials Science and Engineering,
Massachusetts Institute of Technology

**Abstract**
Welding is an extremely complex process; however, due to its commercial importance, it is essential that a more thorough study of the various processes be undertaken. Three examples of areas requiring greater understanding of arc physics are presented. The first discusses heating or cooling of the metal at the cathode; the second describes variations in heat transfer using various shielding gases and the last describes droplet formation in gas metal arc welding.
Key Words: Arc Welding, Arc Physics, Shielding Gases, Gas Metal Arc Welding.

## 1. Introduction

Langmuir, Compton and Cobine are but a few of a number of outstanding physicists who contributed greatly to our understanding of atmospheric pressure arcs during the first half of this century. However, during the past forty years, few physicists have studied such problems. I have been told by some physicists that arc physics is too mundane and too simple a problem to challenge a modern day physicist. This, of course, is merely an excuse. Atmospheric pressure arcs are too complex and represent too difficult a problem for these people to study. Stated simply, atmospheric pressure arcs and welding present problems which are too difficult and too complicated to be approached by the type of physicists that are educated in today's world.

The complexity of welding is readily apparent when one considers that fusion welding involves temperature gradients of thousands of degrees, over distances of less than a centimeter, occuring on a time scale of seconds, involving multiple phases of solids, liquids, gases and plasma. This complexity causes most scientists to shun welding as unscientific and unworthy of thoughtful analysis. It is only the true engineer who is willing to deal with a process of such exceeding complexity in order to achieve the end result of a fabricated product of commercial and societal usefulness. One of the challenges facing those who engineer welded structures and components, is communication of the importance and significance of welding to other scientists and to the general public. One of the purposes of this paper is to describe some of the complexity of several welding processes, and to demonstrate how study of arc physics can help provide a better understanding of arc welding processes.

## 2. Anode and Cathode Spot Heating

Cobine [1955] and others have shown that the heat transferred to the anode, $Q_A$, is given by

$$Q_A = I\left(\phi_A + V_A + \frac{3}{2}\frac{kT}{e}\right)$$

where: $I$ is the welding current in amperes
$\phi_A$ is the work function or heat of condensation of the electrons expressed in volts
$V_A$ is the anode voltage drop
$k$ is the Boltzmann's constant
$T$ is the absolute temperature of electrons in the plasma, and
$e$ is the electron charge

By symmetry, the heat at the cathode should be given by

$$Q_C = -I\left(\phi_C + V_C + \frac{3}{2}\frac{kT}{e}\right)$$

where the sign of the current is reversed due to reversal of electron flow. Such a simplistic formulation of cathode heating may not be valid due to the much more complex (and essentially unknown) physics which occur in the cathode spot [Ecker, 1980]. Nonetheless, the sign reversal of the current clearly indicates that the exiting electrons cool the cathode. The removal of electrons to the plasma causes a cooling of the cathode even as a vaporizing gas removes heat from a liquid.

Experimental confirmation of cathodic cooling was obtained by Ishikawa et al. [1965] who used an infrared pyrometer to show that the hottest portion of a gas tungsten welding arc is not located at the tip, but several millimeters up the electrode. When the cathode is liquid, as in gas metal arc welding, convection within the weld pool essentially negates the local cooling effect of electron vaporization. Thus, cathodic cooling is an important factor when the cathode is solid; but when the cathode is liquid, cathodic cooling has little influence on temperature gradients, although it may alter the overall heat balance.

Nonetheless, cathodic cooling may be important in influencing cathode spot motion on the liquid weld pool. Electron emission from the weld pool is influenced by the temperature of the cathode; a hotter cathode produces emission more readily. When gas metal arc welding of most metals, the cathode spot wanders rapidly across the liquid surface. Does it do this because cathodic cooling produces easier electron emission in the hotter areas adjacent to the cathode spot? Or does cathodic cleaning of surface active oxygen at the cathode spot create a region of easier electron emission on adjacent surface regions where a thin oxygen film remains? Does the addition of oxygen to argon when gas metal arc welding of steel produce an oxygen surface film that influences arc stability? What is the effect of metal vaporization on the properties of the plasma immediately above the cathode spot? Does this high concentration of easily ionized metal vapors create stability for the cathode spot or does it cause the spot to wander about the cathode surface? These and other questions remain unanswered, yet our process modelling tools have progressed to a sufficient level that some could be answered - at least qualitatively.

Clearly, there is much to be learned about the science of atmospheric pressure arcs as they relate to welding. Some of these questions could lead to answers of commercial importance.

## 3. Heat Transfer Using Various Shielding Gases

It is well known that one will melt more metal with a 100 ampere helium arc than with a 100 ampere argon arc. The conventional wisdom says that this is due to the higher ionization potential of helium (24.6 eV) as compared to argon (15.8 eV). It is assumed based upon the Saha equation that this will result in a higher arc temperature. Unfortunately, based on experimental measurements [Key, 1983], there is not a higher temperature in the helium welding arc; and even if there were, it could not explain the much greater melting efficiency when using helium. The correct explanation lies in the variation of the thermal conductivity of the gases.

Although eighty percent of the heat in the argon arc is carried by the electron current, twenty percent is due to heat conductance across a gas boundary layer at the anode surface. [Quiqley et al., 1973; Dinesclu and Pfender, 1980] Heat flow across this boundary layer is controlled by the thermal conductivity of the gas. For monatomic gases such as helium and argon, the thermal conductivity is controlled by the mass diffusivity of the atoms, which, according to the kinetic theory of gases, is proportional to the inverse square root of the mass of the atom [Dunn and Eagar, 1986]. Thus, with an atomic weight ten times as great as helium, argon has a thermal conductivity only 30 percent as large as helium. All other factors being equal, helium will conduct three times as much heat across the boundary layer as will argon. Table 1 illustrates this point. Although only 20 percent of the heat entering the anode from the argon arc is conducted across the boundary layer, 43% of the heat in a helium arc will be conducted across the boundary layer. (These figures are only "back of the envelope" calculations. Many other parameters change when switching from argon to helium gas. Other factors such as anode voltage drop, electric current distribution, arc temperature or radiation will change, but by insignificant amounts as compared to the thermal conductivity.)

As pointed out by Giedt et al. [1989], the actual melting efficiency of the arc welding process is relatively low (i.e. on the order of 20 percent or less). Thus an increase of 40% in the heat supplied could double the volume of metal melted. This is why helium shielded arcs melt more metal than argon shielded arcs.

| Shielding Gas | Electric Current | Conduction across gas boundary layer | Percent by conduction |
|---|---|---|---|
| Argon | 80% | 20% | 20% |
| Helium | 80% | 60% | 43% |

Table 1. Relative Heat Transfer from the Welding Plasma in Argon and Helium Shielding Gases (All heat values normalized to 100% for argon).

In addition to these pure gases, this hypothesis can explain why gas mixtures such as 75 Ar - 25 He or 50 Ar - 50 He will increase the rate of heat transfer over that of pure argon. Such mixtures have a thermal conductivity intermediate to the two pure gases. The ionization potential theory completely fails to explain why argon rich mixtures increase the rate of heat transfer. Since only 5 to 30% of the arc need be ionized, gas mixtures of less than 70 percent He contain no He ions. [Kim, 1986] If the He does not ionize, its ionization cannot be used to explain the more efficient heat transfer.

One may also ask why Argon-5% $H_2$ shielding gases also transfer heat more efficiently than pure argon. Using the simple mass diffusivity, thermal conductivity relationships presented above with a simple rule of mixtures, one would conclude that Ar-5% $H_2$ would have approximately a 25% greater thermal conductivity than pure argon. This cannot explain the greater rate of melting observed as this increase in thermal conductivity would result in only a 5% increase in the amount of heat transformed to the anode. The explanation in this case is due to a third mechanism of heat conduction. In addition to the flow of the electric current and the thermal conductivity across the gas boundary layer, diatomic gases such as hydrogen also have an enhanced thermal conductivity due to dissociation of the gas in the plasma and recombination in the gas boundary layer. This enhancement of thermal conductivity in polyatomic gas plasmas is known as the reactive thermal conductivity. [Dunn, 1984] In fact, the thermal conductivity of Ar-5% $H_2$ is enhanced even more than 25% over that of pure argon, due to this contribution of the reactive thermal conductivity.

## 4. Droplet Formation in GMAW

Photography of the gas metal arc process shows that the anode spot corona varies markedly with different shielding gases. With helium or carbon dioxide shielding, the anode spot is attached to the bottom of the pendant drop on the end of the electrode. The size of the spot does not change markedly as the current increases. With argon, the anode spot is more diffuse and grows noticably in size as the current increases. At sufficiently high currents (for example, above 230 amperes with 1.6 mm diameter steel electrodes) the anode spot fills the molten drop and begins to climb the cylindrical side walls of the electrode. Since most of the heat carried to the anode is carried by the welding current, the current in the anode spot causes the cylindrical side walls of the electrode to melt [Y.S. Kim, 1989]. The end of the electrode forms a tapered geometry as shown in Figure 1. Such a tapered electrode tip geometry does not develop in pure helium or pure carbon dioxide shielding gases.

The tapered tip creates a smaller diameter for attachment of the droplet to the electrode by surface tension forces [Y.S. Kim, 1989]. This produces smaller droplets than would be present in the absence of taper formation. The smaller droplets formed with argon shielding from a tapered electrode are seen clearly in Figure 2, as compared with helium and carbon dioxide shielding.

The formation of the taper occurs during gas metal arc welding of steel, aluminum and titanium, when using argon shielding gas [Y.S. Kim, 1989]. There are a number of consequences of this taper formation. Firstly, Y.S. Kim [1989] has shown that the melting rate of the electrode is controlled by heat transfer across the liquid-solid

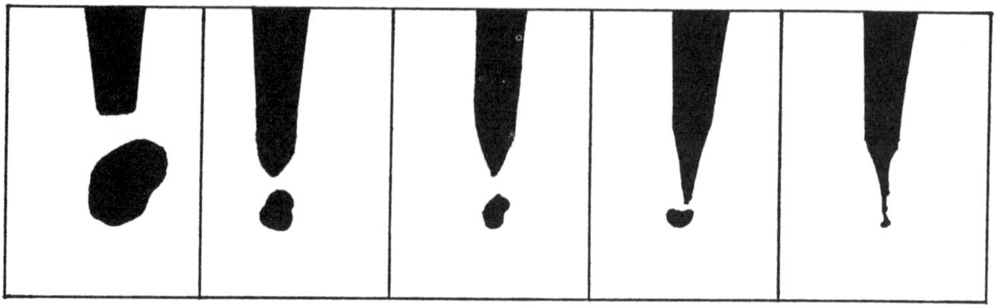

Fig. 1. Development of a taper as welding current increases when using argon shielding gas. (a) 205 amperes, (b) 237 amperes, (c) 253 amperes, (d) 281 amperes, and (e) 310 amperes.

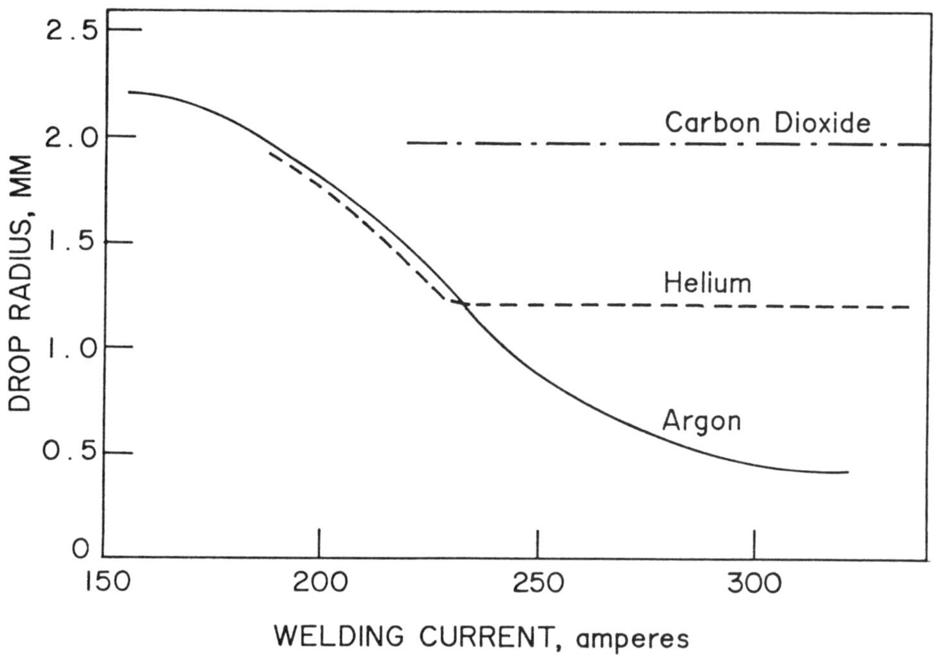


Fig. 2. Droplet size measured in three different shielding gases as a function of welding current for 1.6 mm diameter steel electrodes.

boundary between the drop and the solid electrode. Since the taper reduces the area of this boundary, taper formation reduces the rate of melting of the electrode. At high currents, the melting rate in helium and carbon dioxide is greater than in argon due to absence of the taper. Thus the greater melting rate in helium is not due to the higher ionization potential of this gas as is commonly assumed (see Section 3 of this paper), but to a larger liquid-solid contact area in which heat is transferred to the solid electrode.

Secondly, Y.S. Kim has shown that the formation of a taper can reduce the range of one drop per pulse transfer in pulsed current GMAW. As the drops become smaller with argon shielding gas due to taper formation, the pulsing frequency must increase accordingly. By suppressing the formation of the taper, both large and small drops can be released from the end of the electrode during pulsed current welding. The largest drops are equal to the droplet size created at the low current portion of the pulse in the absence of the taper, while the smallest drops are equivalent to those produced during the peak of the current pulse. This greatly increases the range of frequencies which are permissible for one drop per pulse operation [Y.S. Kim and Eagar]. By addition of helium or carbon dioxide to the argon shielding gas, the anode spot is constricted and does not climb the cylindrical side walls of the electrode as readily. Thus taper formation is suppressed and the acceptable range of frequencies for pulsed current GMAW is broadened and the process becomes more easily controlled. This example illustrates how an understanding of the physics of droplet formation can aid the welding engineer in selection of the best shielding gas composition for pulsed current gas metal arc welding.

## 5. Conclusion

It is seen that an understanding of the arc physics can make a very complex process such as welding, appear much clearer. By understanding the process, it becomes easier to control and easier to modify our procedures to ensure that high quality, reliable joints are produced. With so many arc process variations and with an ever increasing number of materials to be joined, there is ample opportunity for many people to contribute to a better understanding of the physics of welding processes.

## 6. Acknowledgements

The author expresses his appreciation to the U.S. Department of Energy, the Office of Naval Research and the National Science Foundation for support of the research presented herein.

## References


Dinulescu, A. and Pfender, E. (1980) Analysis of the Anode Boundary Layer of High Intensity Arcs. J. Appl. Phys., 51(6), 3149-3157.

Dunn, G.J. (1984) Metal Vapors in Gas Tungsten Welding Arcs. S.M. Thesis, Dept. of Materials Science and Engineering, MIT, Cambridge, MA.

Dunn, G.J. and Eagar, T.W. (1986) Metal Vapors in Gas Tungsten Arcs: Part II. Theoretical Calculations of Transport Properties. Metall. Trans., 17A(10), 1865-1871.

Ecker, G. (1980) Theoretical Aspects of the Vacuum Arc, in Vacuum Arcs, Theory and Application (ed. J.M. Lafferty), John Wiley, NY, pp. 228-320.

Giedt, W.H., Tallerico, L.N. and Fuerschbach, P.W. (1989) GTA Welding Efficiency: Calorimetric and Temperature Field Measurements. Welding J., 68(1), 28s-32s.

Key, J.F., Chan, J.W. and McIlwain, M.E. (1983) Process Variable Influence on Arc Temperature Distribution. Welding J., 62(7), 179s-184s.

Kim, E.W. (1986) Visible Light Emissions During Gas Tungsten Arc Welding. S.M. Thesis, Dept. of Materials Science and Engineering, MIT, Cambridge, MA.

Kim, Y.S. (1989) Metal Transfer in Gas Metal Arc Welding. Ph.D. Thesis, Dept. of Materials Science and Engineering, MIT, Cambridge, MA.

Kim, Y.S. and Eagar, T.W. Metal Transfer in Pulsed Current GMAW. Submitted to Welding J..

Quigley, M.B.C., Richards, P.H., Swift-Hook, D.T. and Gick, A.E.F. (1973) Heat Flow to the Workpiece From a TIG Welding Arc. J. Phys. D. - Appl. Phys., 6, 2250-2258.

Savage, W.F., Strunck, S.S. and Ishikawa, Y. (1965) The Effect of Electrode Geometry in Gas Tungsten-Arc Welding. Welding J., 44(11), 489s-496s.

# HEAT AND FLUID FLOW IN WELDS

J. GOLDAK AND M. J. BIBBY, D. DOWNEY AND M. GU
Mechanical & Aerospace Engineering, Carleton University, Ottawa, Canada

Abstract
The current state and future directions of numerical modeling of heat and fluid flow in welds are presented. Weld pool models have evolved rapidly in the past decade and full 3D analyses are now possible for short welds. In two dimensions, modeling the formation of defects has begun. Outside of the weld pool, computational heat flow analyses are now able to analyze welds of medium length with useful resolution and accuracy. Prescribed flux and power density functions or prescribed temperature functions are used to model the heating effect of the arc. Kinematic models have evolved from 2D cross-sectional to 2D in-plane to full 3D and shell-3D composite models. The coupling of heat flow models to the evolution of microstructure and thermal stress analysis has resolved the effect of transformations on the residual stress in high strength steels. The paper closes with a discussion of future opportunities and challenges.

Key words: Computational Weld mechanics, Weld pool, Heat transfer, Arc welding.

## 1. The Weld Pool

Our understanding of the physics of the weld pool prior to 1980 is clearly explained in a beautiful book written by J. F. Lancaster [1]. The basic physics of surface tension in fusion welding is discussed in detail in the four part paper by Matsunawa [2]. Heiple and Roper's experimental studies strongly influenced weld pool research in the past decade [3]. Specific weld pool models are described in references [4-12].

Prior to 1980, the primary mechanism for heat transfer in weld pools was known to be convection, i.e., heat is transported by the motion of the liquid and not by diffusion. Although the basic forces acting on the weld pool were known to be due to buoyancy, electromagnetic, Marangoni and arc pressure, the relative importance of these forces was not well understood. This question was resolved by the numerical models that began to appear in the 1980s [4-12]. These models not only demonstrated that the effect of buoyancy, electromagnetic, Marangoni and arc pressure forces on the weld pool velocity, pressure and temperature could be computed but also demonstrated that numerical modeling was a powerful tool for understanding the weld pool.

It can be argued that Zacharia, Ersalan, et al, presented the most comprehensive analysis of the weld pool performed to date [10,11]. Their model was fully 3D and transient. The weld pool-arc and liquid-solid interfaces were free boundaries. Filler metal was added to the weld pool from the arc. Computational costs limited the arc travel to a short distance. MacLay and Carey used a finite element formulation for an axisymmetric weld pool with both flat and depressed weld pool surfaces. As more distorted weld pools are studied, the use of FEM is likely to become more important. Kou studied the effect of the normal and shear stress induced on the weld pool surface due to surface tension in a deformed weld pool [7]. Mazumder considered the numerical requirements in some detail [6]. More recently, Szekely [9] and Matsunawa [9] have both studied the effect of sloshing in the weld pool.

Although progress in modeling the weld pool is impressive, the capability of current models to predict weld pool behavior is limited by several factors. The data required include the heat flux, current density distribution and pressure velocity field from the arc; the surface tension as a function of composition and temperature of the weld pool surface; the viscosity, density and vapor pressure of the liquid metal in the weld pool. Those data are difficult to obtain, the accuracy is modest and they cannot yet be predicted for most welding procedures. The computing costs are high. These factors have forced most analyses to be 2D and often stationary. The arc interacts with the weld pool shape and the weld pool shape affects the arc. Therefore it will be important to study not only the weld pool but the weld pool and arc as a single coupled system. This is particularly important for weld pools with deep depressions. Szekely has made a useful start on coupling the arc and weld pool [9]. The primary contribution of weld pool models to date is in understanding the behavior of weld pools and the importance of various factors. Although it will be some time before weld pool models realize their ultimate dream; to predict the weld pool shape for a given welding procedure, there is little doubt that to truly understand, control or predict welds and welding processes, we must model weld pools accurately and efficiently.

Remark: Because of limited space only the most critical or most recent references are cited to provide the reader with an entry into the literature. We apologize to authors of references that have been omitted.

## 2. Transient Heat Affected Zone

### 2.1 Background

The classical analysis of heat flow in welding was pioneered by Rykalin and Rosenthal [13, 14]. Rykalin's Houdremont lecture of 1972 and the review paper by Myers et al [15] summarize the research from this classical viewpoint. This paper focuses on numerical methods which have dominated the study of heat flow in welds since Westby's pioneering analysis of 1965 [16]. The major research issues are; i) modeling the distribution of heat input from the arc; ii) the kinematic model - 2D cross-sectional, 2D in-plane, 3D or 3D-shell composite model; iii) the design of FEM meshes; iv) the choice of solver and time integration scheme; v) the computational cost and accuracy obtained; vi) the study of errors; vii) the possibilities of obtaining optimal solutions; viii) coupling heat transfer to microstructure, thermal stress analysis including transformation plasticity; ix) unsolved problems including the problem of multi-pass welds.

### 2.2 Theory

The conservation of energy is the fundamental principle that underlies the analysis of heat flow. It is usually expressed in the form of a partial differential equation which can be found in any text on heat transfer. A finite difference approximation to the derivatives leads to the finite difference method. The error depends on the difference between the finite difference approximation and the derivative. However, finite element analysis, FEM, is based on an integral formulation which minimizes the error in a least square sense, i.e., the error is made orthogonal to the test function space. The complete statement of a well posed heat flow problem requires the domain in space (geometry), the boundary conditions, initial conditions and the thermal properties be specified and consistent.

2.3 Modeling the Heating Effect of the Arc

It is not difficult to model the geometry, initial conditions, the total heat input or to specify the thermal properties. The difficult tasks are to measure the thermal properties and to estimate the distribution of energy in the weld pool. If the energy equation is coupled to the weld pool physics, the energy distribution would be determined from the weld pool physics. Although this is the proper method for determining the energy distribution, solutions to this coupled problem have not yet been demonstrated.

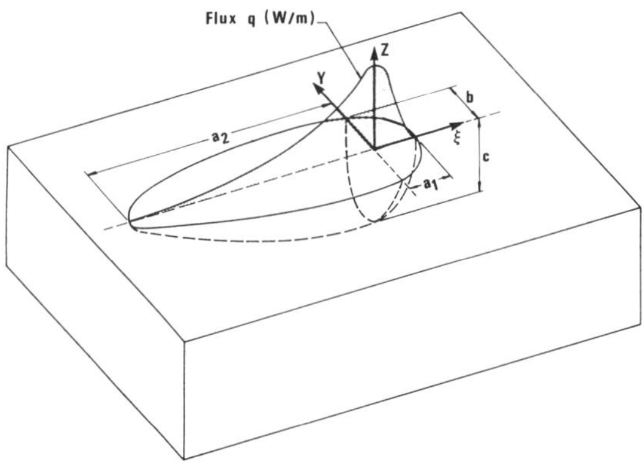


Figure 1. The heating effect of the arc is modeled with a distribution of flux over the surface and the power density throughout the volume of the double ellipsoid.

If the uncoupled energy equation is to be solved, there are two fundamental options. Either the thermal flux and power density distribution can be specified, i.e., natural boundary conditions, or the temperature can be specified, i.e., essential boundary conditions. Of course, an infinite number of functions can be chosen for either option. In general, the greatest success has been obtained by constraining the heat input to the region of the weld pool. This means that heat transfer is not used to predict the shape and size of the weld pool but only to predict the temperature field outside of the experimentally determined or estimated weld pool. Having restricted the heat input to the weld pool, the next question is how to distribute the heat input in the weld pool. The simplest function is a constant distribution. Andersson used a piece-wise constant function [17]. The next simplest is a Gaussian distribution function [18-20]. The simplest shape is a circle in 2D or a sphere in 3D. The next simplest shapes are ellipses or double ellipses in 2D or ellipsoids or double ellipsoids in 3D. Almost all research to date has chosen to specify the flux or power density distribution. This research is summarized in references [19-20].

Although the power density models were able to analyze non-spherical weld pools, such as electron beam and MIG welds with nail heads [20,25], attempts to analyze welds with complex pools such as a double pool achieved limited success. In addition, recent advances in experimental techniques enable the temperature distribution of the weld pool to be measured accurately [21]. These factors have motivated the study of prescribed temperature heat source models. The prescribed temperature model is only valid after the weld pool has reached steady state. This is typically two to five weld pool diameters from the arc starting point. In this region it has the potential to analyze more complex weld pool shapes (Figure 2). In addition, the distributed flux and power density computed from the prescribed temperature model can be applied to analyze the starting and ending transients.

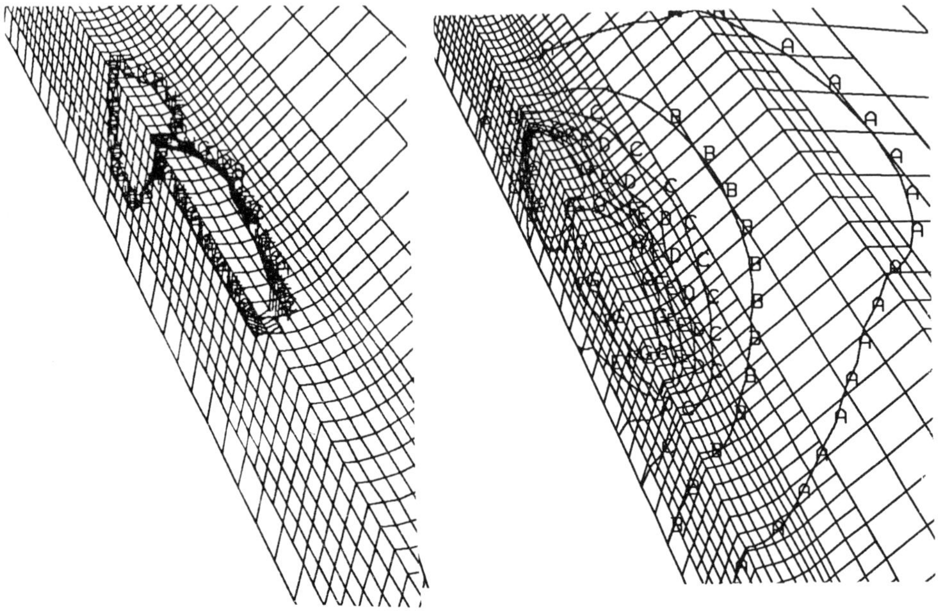

Figure 2. A thermal analysis of a weld using a prescribed temperature weld pool model. The prescribed weld pool temperature distribution is shown on the left. Note the deep pool under the arc and the wider shallower pool behind it in this complex two pool model. The contours of the computed temperature field are shown on the right.

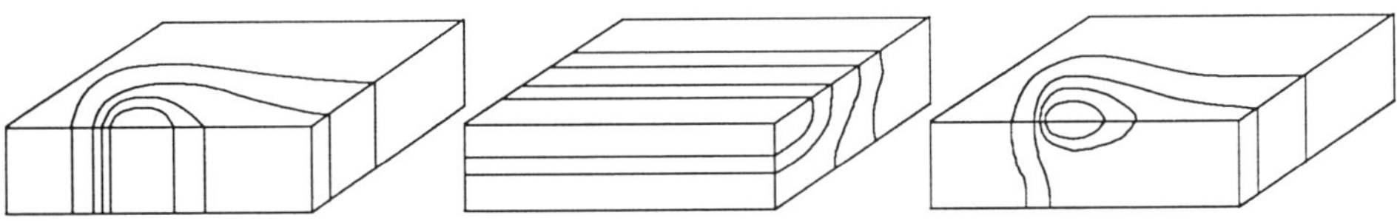

Fig. 3. By constraining the thermal flux and hence the temperature gradients, the kinematic models are consistent with the temperature fields shown in; a) in-plane, b) cross-sectional, c) 3D model.

2.4 Kinematic Models

Although the reality is that heat flow is 3D and transient, analysts have long sought lower dimensional models to minimize computing costs. The options are illustrated in fig. 3. The most popular has been the cross-sectional model which assumes that heat flow in the direction of welding is zero. This is a surprisingly subtle model. Strictly, it is limited to steady state analysis of prismatic geometries with the weld parallel to the prismatic axis. How can such a model possibly be accurate since some heat clearly must flow in the direction of welding? The answer is that the heat input has been modified to account for the heat that flows in the longitudinal direction. Because it uses relatively few 2D elements, it is computationally one of the cheapest models.

The next most popular kinematic model has been the in-plane model. It assumes the heat flow normal to the weld plate is zero. The advantage of the in-plane over the cross-sectional model is its capability for computing the starting and ending transients. However, because it tends to require rather more elements and more time steps, the computational costs are higher [22]. It has not be used extensively.

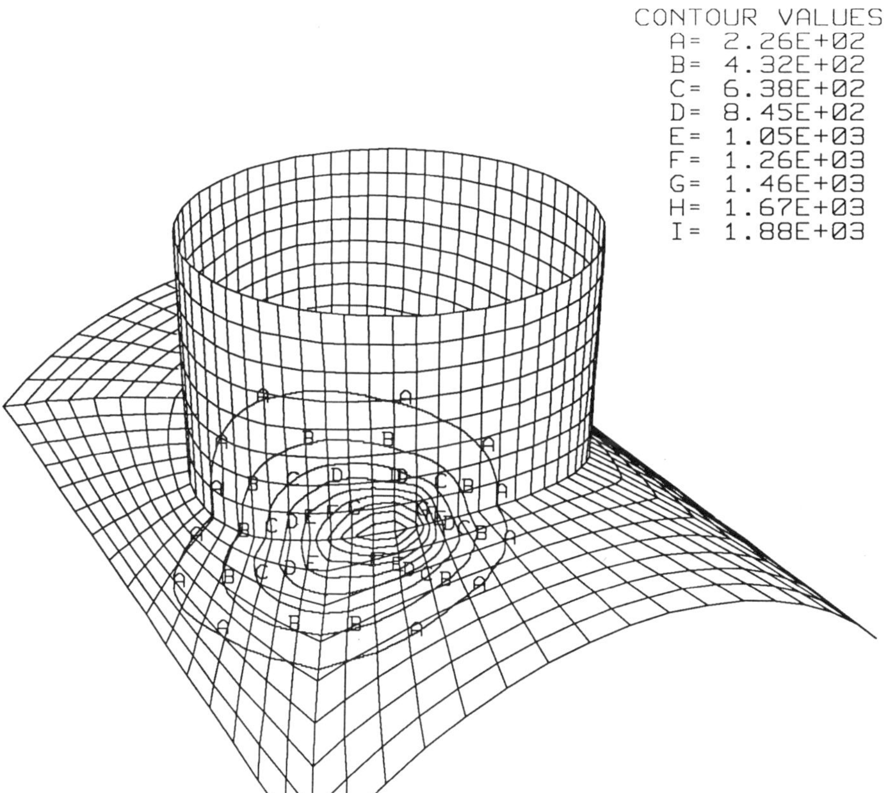


Figure 4. The contours of the transient temperature field is shown for a weld of a tee joint. The pipe is 40 mm in diameter and the wall thickness is 2.5 mm. Thermal shell elements are used to reduce the computing cost. The welding parameters are voltage 32, amperes 150, speed 1 mm/s, power input 1.5 Kj/mm.

To the author's knowledge, the use of thermal shell models in welds have not been published to date. The thermal shell is the generalization of the in-plane model to curved surfaces. It assumes that no heat flows normal to the plane and that all heat flows in the tangent plane, i.e., the temperature gradient lies in the tangent plane. Where these conditions hold to an adequate accuracy, the shell model has several advantages. The shell element usually has half the number of degrees of freedom and half the bandwidth. This reduces computing costs by a factor of eight for direct solvers. Even more important, it allows larger elements than the equivalent 3D elements because they are better conditioned numerically. Roughly speaking, the shell element becomes more accurate the thinner the plate and the farther one moves from the welding arc. An example is shown in figure 4.

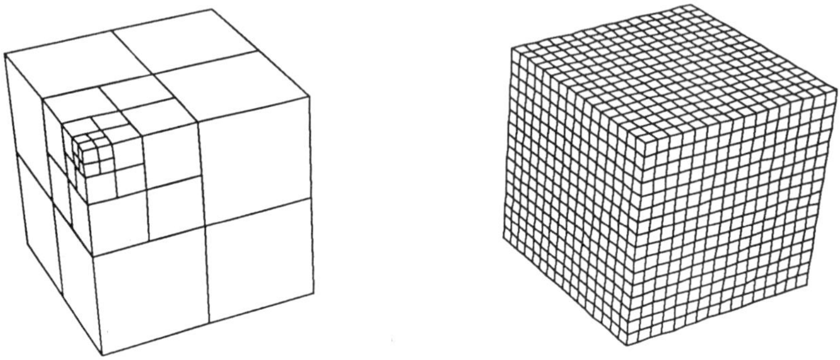

Figure 5. The mesh created with graded elements in a) is just as accurate as the uniform mesh in b) for a problem with a point load at the front, left, top corner but the computing cost is dramatically reduced. Mesh a) has 29 elements while mesh b) has 4096 elements.

Because reality is 3D and transient, and 3D analysis is the most general formulation, analysts have long sought to perform full 3D analysis of welds. The first analyses were reported in 1986 [23-25]. In 3D analyses the kinematic model is obviously correct. The issue is to reduce the computational cost for meshes that achieve useful accuracy to acceptable levels. The principal weapons are optimizing the mesh and creating efficient solvers.

Here, the grading elements of McDill have made a major contribution in optimizing the mesh for the analysis of welds [26]. These elements reduce the computational complexity from $O(n^7)$ for a cube with uniform mesh with n elements on an edge to O(log n) for the cube with a graded mesh. In both cases, the problem has a point load on one corner of the cube and direct solvers are used. In this problem, there is no loss of accuracy with the graded mesh.

The second major weapon is the use of improved solvers. The use of conjugate gradient solvers with element-by-element preconditioning has produced dramatic gains. Because these solvers do not assemble or invert the global stiffness matrix, the memory requirements are greatly reduced. The transient temperature fields shown in figures 2 and 4 were computed on a workstation with such a solver. Typical computing times are 120 seconds per time step for a mesh with 10,000 nodes on a 0.9 mflop workstation.

Since the 3D model is needed near the arc and the shell model is most efficient far from the arc, can these models be combined to produce a more efficient model? Fig. 7 shows such a composite 3D-shell model. It requires a special transition element to join shell and 3D brick elements.

2.5 Coupling Heat Flow to Microstructure and Thermal Stress
Since welding engineers are interested in heat flow because it drives and largely controls the microstructure, thermal strain and stress in welds, it is not surprising that analysts have sought to model heat flow coupled to microstructure and thermal stress.

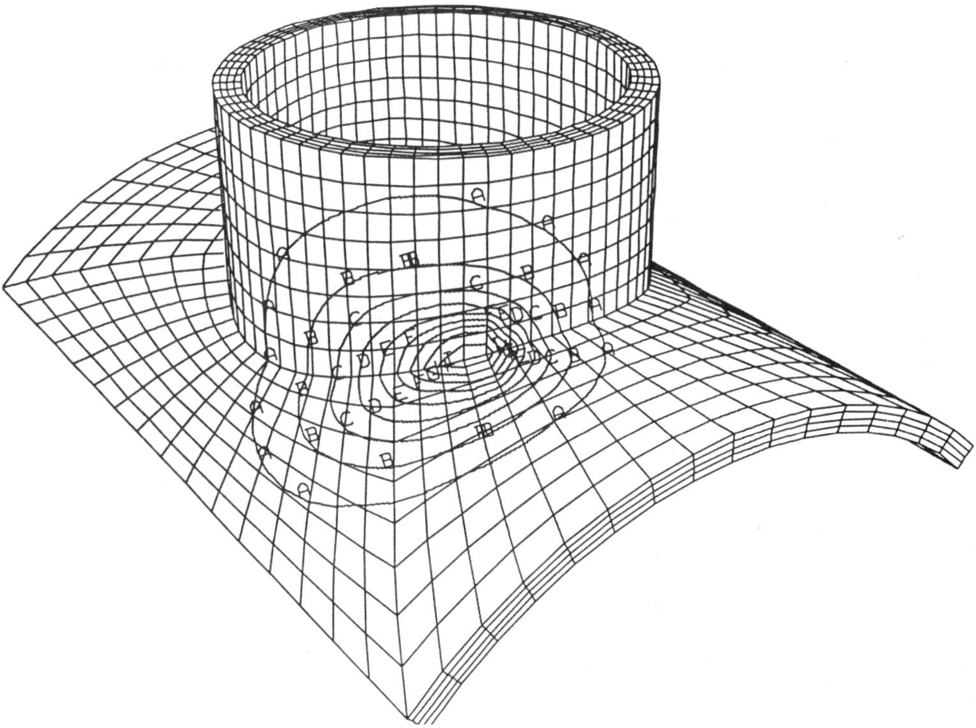

Figure 6. This tee joint is modeled with 8 noded 3D brick elements. This is the most accurate and the most expensive kinematic model.

For low alloy steels, the evolution of microstructure in the HAZ can now be computed efficiently [28-29]. Basically phase transformations are assumed to occur at equilibrium on heating with no superheating. This analysis uses the iron-carbon phase diagram as a function of some fifteen alloying elements. Ordinary differential equations for the grain growth of austenite and the decomposition of austenite to ferrite, pearlite and bainite are integrated over time (figure 8). The computation of microstructure in the FZ is not yet coupled to numerical thermal analysis [30]. Although the effect of latent heat from phase changes on the temperature field has relied primarily on simple specific heat and enthalpy models, recent research is turning to more sophisticated models for capturing latent heat effects [31].

The thermal stress analysis of welds is more complex than the heat flow analysis because the geometry changes and because of the complex stress-strain relationship. Most thermal stress analyses performed to date have used thermo-elasto-plastic constitutive models. Since this model is certainly not valid in the liquid region and is suspect near the melting point where viscous effects are expected to be important, most analyses assume a cutoff temperature. They assume that the thermal strain, Young's modulus and yield strength do not change above a cutoff temperature. For steels the cutoff temperatures used have ranged from 600 C to 1200 C. Computed displacements in a weld are shown in Figure 9. Although the accuracy is limited by computational costs to rather coarse meshes, the residual stresses have usually agreed well with experimental results.

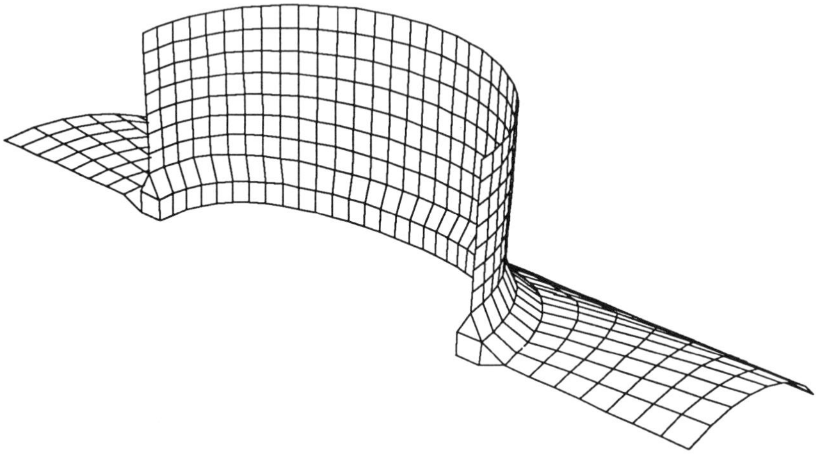

Figure 7. This tee joint is modeled with 4 noded thermal shell elements far from the weld joint, 8 noded 3D brick elements at the weld joint and special transition elements to join the shell and brick elements. This composite kinematic model is more accurate than the shell mesh shown in Figure 4 and computationally cheaper than the mesh using 3D bricks shown in Figure 6.

An important exception has been the residual stress in high strength steels. Hibbitt and Marcal [33] attempted to predict the residual stress in welding HY-130 steel. Their results were so different from experiment that they concluded that an important physical phenomena had been missed (Figure 10). In 1988, A. Oddy et al repeated the analysis but included the effect of the austenite to martensite transformation [34]. They obtained the correct residual stress and showed that the critical phenomena missed by Hibbitt and Marcal was transformation plasticity. This is the plastic strain that occurs during a phase change in the presence of a deviatoric stress.

## 2.6 Numerical Solutions

Although finite difference and finite volume methods dominate fluid flow analysis, heat flow and thermal stress analysis are dominated by finite element methods. Finite difference, finite volume, and boundary methods will not be discussed here. The finite element method is a form of functional analysis. Two function spaces are chosen. One is an infinite dimensional space which contains the exact solution to all properly posed problems in a given class and the other is a finite dimensional space on which exact solutions are projected. Most importantly, all exact solutions in this class can be projected onto the finite dimensional subspace. The error is orthogonal to the subspace, i.e., FEM provides the most accurate solution possible for that subspace.

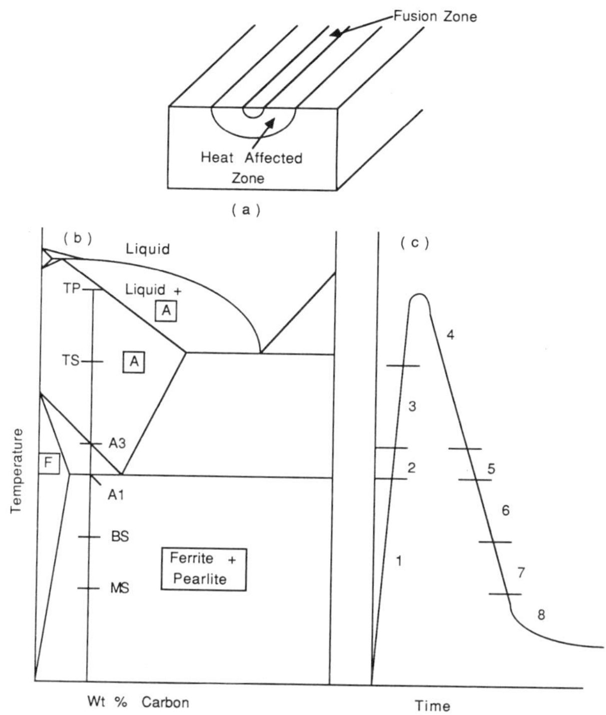


Figure 8. The microstructure at a point in the HAZ shown in a) under goes the thermal cycle shown in c). The microstructure evolves through the following stages: 1) ferrite and pearlite are at equilibrium; 2) ferrite and pearlite transform to austenite; 3) austenite grain growth is inhibited by NbC and VC; 4) austenite grain growth occurs; 5) austenite decomposes to ferrite; 6) austenite decomposes to pearlite; 7) austenite decomposes to bainite; 8) austenite decomposes to martensite.

## 3. Future Developments

There is every reason to expect computational weld mechanics to develop at an ever increasing rate for the next decade. The increase in performance of workstations that cost roughly $30,000 is shown in table 1. In the past four years, Mflops have increased 20 times. The MIPS 6000, Intel i860 and Motorola 68040 microprocessors which are expected to be released shortly will continue this trend. Certainly, researchers have every reason to expect to have 100 Mflops workstation on their desks before 1995 at costs of less than $50,000. Clearly, the cost of hardware per cpu cycle is rapidly falling to zero.

The rapid improvement in algorithms is just as important as faster, cheaper hardware. In the authors' laboratory in 1985, one time step of a 1600 dof problem required 16 hours of cpu time. The same problem runs in 10 minutes today with the same computer. The factor of 100 is due only to software. On a problem with 6,000 dofs, the computing time has fallen by a factor of 6 in the past year. Last year problems with more than 15,000 equations could not be computed. In 1989 problems with 42,000 equations were computed at rates of approximately 15 minutes per time step.

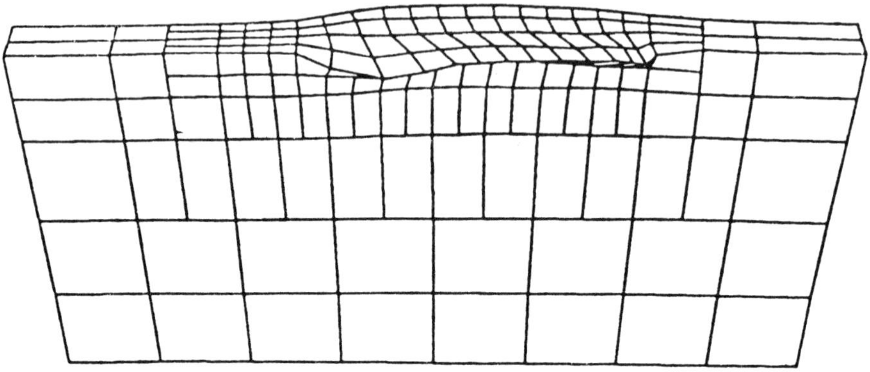

Figure 9. Displacements computed by a thermo-elasto-plastic analysis for a 1.6 KJ/mm edge weld on a carbon steel bar 12.7 mm in width. Note that the displacements have been magnified by a factor of 50 to make them discernable. Note the displacements in the longitudinal direction that violate the assumptions of plane strain.

Table 1. The evolution of microprocessor performance since 1986

| Microprocessor | MHz | Mips | Mflops | Memory | Year |
|---|---|---|---|---|---|
| Motorola 68020 | 12 | 1.0 | 0.08 | 1.5 | 1986 |
| Motorola 68020 | 25 | 3.0 | 0.16 | 8 | 1987 |
| Motorola 68030 | 25 | 4.0 | 0.32 | 32 | 1989 |
| MIPS 2000 | 12 | 11 | 0.9 | 32 | 1988 |
| MIPS 3000 | 20 | 20 | 1.6 | 32 | 1989 |
| MIPS 6000 | | | | | |

Although improvements in hardware and algorithms will enable weld models to be analyzed quickly and cheaply, researchers face major challenges in the next decade. The analysis of multi-pass welds may be the most important. Such models should be coupled to phenomena such as hydrogen cracking, fracture mechanics, continuum damage mechanics and weld defects. The models should be verified and coupled closely with experiments. They should be used to design experiments and the experiments used to verify the models. User interfaces are needed to allow welding engineers to use these models without becoming experts in FEM analysis, computer science or metal physics. As surely as we expect the sun to rise tomorrow, we expect these developments to be realized in the very near future.

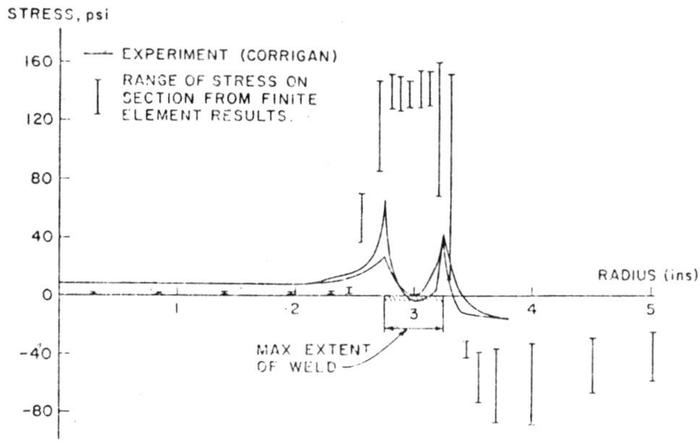


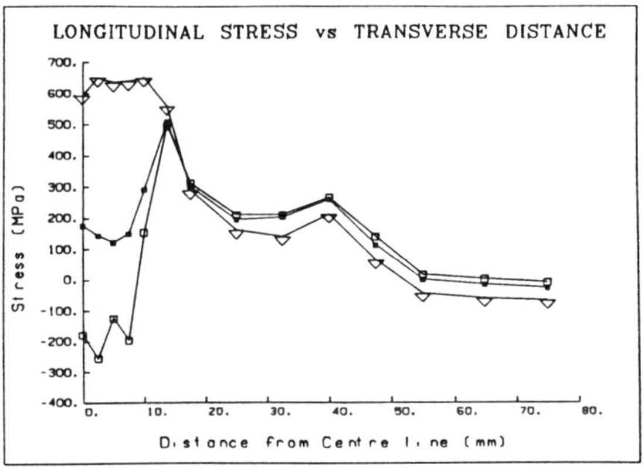


Fig. 10 The longitudinal stress in HY-130 weld measured by Corrigan and computed by Hibbitt & Marcal are shown in a). Note the lack of agreement between computed and measured stresses. The longitudinal stress shown in b) was computed by Oddy et al [34] for HY-80 for three cases. The top curve ignores the phase transformation. The bottom curve includes the volume change due to the transformation. The middle curve includes the effect of transformation plasticity. Note the agreement between experiment and computed results.

References

1. J. F. Lancaster, The Physics of Welding, 2nd Edition, Pergamon Press, 1986.

2. A. Matsunawa, Role of Surface Tension in Fusion Welding, IIW Doc. 212-618-85

3. Heiple, C. R. Roper, J.R., Stagner, R.T., and Alsen, J.J. 1983. Surface active element effects on the shape of

GTA, laser and electron beam welds. Welding Journal 62(3):72-s

4. Athey;, D.R. 1980. A mathematical model for fluid flow phenomena in weld pools. Journal of Fluid Mechanics. Vol.98, p.787.

5. Oreper, G.M., and Szekely, J. 1984. Heat and fluid flow phenomena in weld pools. Journal of Fluid Mechanics. Vol. 147, p 53.

6. Zehr, R., Chen, M.M. and Mazumder, J. Generic Structure of Thermocapillary Flow in Weld Pools and Its Influence on the Computational Strategy. Modeling and Control of Casting and Welding Processes IV, Ed. A. F. Giamei, G. J. Abbaschian, TMS, Palm Coast, Florida, April 17-22, 1988, pp 433-446.

7. Tsai, M. C. and Kou, S. The Advantages of Orthogonal Curvilinear Coordinates in Simulating Marangoni Convection is Deformed Weld Pools. Modeling and Control of Casting and Welding Processes IV, Ed. A. F. Giamei, G. J. Abbaschian, TMS, Palm Coast, Florida, April 17-22, 1988, pp 409-420.

8. R. McLay and G. F. Carey, Coupled Heat Transfer and Viscous Flow, and Magnetic Effects in Weld Pool Analysis, IJNME, Vol. 9,m pp. 713-730, 1989.

9. Proceedings of an International Conference on Trends in Welding Technology, Gatlinburg, Tennesse, USA, ed. S. A. David, May 15-19, 1989.

10. T. Zacharia, A.H. Eraslan, D. K. Aidun, Welder: A Computer Code for Simulating Welding, pp. 177-185, Modeling and Control of Casting and Welding Processes IV, Ed. A. F. Giamei, G. J. Abbaschian, TMS, Palm Coast, Florida, April 17-22, 1988.

11. T. Zacharia, A. H. Eraslan, D.K. Aidun, Modelling of Non-Autogenous Welding, Welding Research Supplement, Welding J., pp. 18-2 to 27s, (1988).

12. E. Prado and D. C. Weckman, Prediction of Weld Pool and Reinforcement Dimensions of GMA Welds Using a Finite Element Model, to appear in Met. Trans B.

13. R. R. Rykalin, Energy Sources for Welding, Welding in the World, 1974, Vol. 12, No. 9/10, p. 227-248, (Houdrement Lecture, International Institute of Welding, London, 1974, p. 1-23).

14. D. Rosenthal, The Theory of Moving Sources of Heat and Its Application tc Metal Treatments, Trans ASME, 1946, vol. 68, p. 849-865.

15. P.S. Myers, O.A. Uyehara, and G.L. Borman, Fundamentals of Heat Flow in Welding, Welding Research Council Bulletin, No.123, New York, 1967.

16. O. Westby, Temperature Distributior in the Workpiece by Welding, Department of Metallurgy and Metals Working, The Technical University, Trondheim, Norway, 1968.

17. Andersson, B.A.B., Thermal Stresses in a Submerged-Arc Welded Joint Considering Phase Transformations, J. Eng. & Tech., Trans. ASME, Vol. 100, 1978, pp. 356-362.

18. V. Pavelic, R. Tanbakuchi, O. A. Uyehara and P.S. Myers, Experimental and Computed Temperature Histories in Gas Tungsten-Arc Welding Thin Plates, Welding Jour-

nal Research Supplement, 1969, vol. 48, p. 295s-305s.

19. J. A. Goldak, A. Chakravarti and M. J.. Bibby, A New Finite Element Model for Welding Heat Sources, Trans. AIME., vol. 15B, 299-305, June,1984.

20. J. A. Goldak, M.J. Bibby, J. Moore, R. House, and B. Patel, Computer Modeling of Heat Flow in Welds, Trans. AIME, Vol. 17B, pp. 587-600, Sept. 1986,

21. H. G. Kraus, Experimental Measurement of Stationary SS 304, SS 316L, and 8630 GTA Weld Pool Surface, Welding Journal, 1989.

22. K. W. Mahin, S. MacEwen, W. S. Winters, W. Mason, M. Kanouff, and E. A. Fuchs, Evaluation of Residual Stress Distributions in a Travelling GTA Weld Using Finite Element and Experimental Techniques, Proc. of Modeling of Casting and Welding Processes IV, Ed. A. F. Giamei and G. J. Abbaschian, pp. 339-350, April 17-22, 1988, Pub. Eng. Foundation and TMS/AIME.

23. K. Mahin, A. B. Shapiro, J. Hallquist, Assessment of Boundary Conditions Limitations on the Development of a General Computer Model for Fusion Welding, pp. 215-224, Proceedings of an International Conference on Trends in Welding Technology, Gatlinburg, Tennesse, USA, May 18-22, 1986, ed. S. A. David.

24. P. Tekriwal, J. Mazumder, Finite Element Modeling of the Arc Welding Process, pp. 71-80, Proceedings of an International Conference on Trends in Welding Technology, Gatlinburg, Tennesse, USA, May 18-22, 1986, ed. S. A. David.

25. J. A. Goldak, M. McDill, A. Oddy, R. House, X. Chi, M. J. Bibby, Computational Heat Transfer for Weld Mechanics, pp. 15-20, Proceedings of an International Conference on Trends in Welding Technology, Gatlinburg, Tennesse, USA, May 18-22, 1986, ed. S. A. David.

26. M. J. McDill, J. A. Goldak, A. Oddy, M. J. Bibby, Isoparametric Quadrilaterals and Hexahedrons for Mesh-Grading Elements, Comm. Applied Nu. Methods, Vol. 3, pp. 155-163 (1987

27. J. Goldak and M. Bibby, Computational Analysis of Heat Transfer of Welds, Proc. of Modeling of Casting and Welding Processes IV, Ed. A. F. Giamei and G. J. Abbaschian, pp. 153-166, April 17-22, 1988, Pub. Eng. Foundation and TMS/AIME.

28. Watt, D.F., Coon, L., Bibby, M.J., Goldak, J.A. and Henwood, C., Modelling Microstructural Development in Weld Heat-Affected Zones, Acta Metal., pp. 3029-3035, Vol 36, No. 11. 1988.

29. Henwood, C., Bibby, M.J., Goldak, J.A. and Watt, D.F., Coupled Transient Heat Transfer-Microstructure Weld Computations, Acta Metal., pp. 3037-3046, Vol 36, No. 11, 1988.

30. H. Bhadeshia, Modelling the Microstructure in the Fusion Zone of Steel Weld Deposits, Proceedings of an International Conference on Trends in Welding Technology, Gatlinburg, Tennesse, USA, ed. S. A. David, May 15-19, 1989.

31. M. Rappaz, S. A. David, L.A. Boatner, and J. Vitek, Development of Microstructures in Fe-15Ni-15Cr Single-Crystal E. Beam Welds, Metall. Trans., to be published.

32. Goldak, J. A., Modelling Thermal Stresses and Distortion in Welds, to be published, Proceedings of an International Conference on Trends in Welding Technology, Gatlinburg, Tennesse, USA, ed. S. A. David, May 15-19, 1989.

33. Hibbitt, H.D., Marcal, P. V., A Numerical Thermo-Mechanical Model for the Welding and Subsequent Loading of a Fabricated Structure, Comp. & Struct., Vol. 3, pp. 1145-1174.

34. A. Oddy, J. A. Goldak, J. M. J. McDill, Transformation Effects in the 3D Finite Element Analysis of Welds, Proceedings of the 2nd International Conference on Trends in Welding Research, Gatlinburg, Tennessee, May 15-19, 1989.

35. R. I. Karlsson and B. L. Josefson, Three Dimensional Finite Element Analysis of Temperature and Stresses in Single-Pass Butt Welded Pipe, accepted for publication, J. Pressure Vessel Technology, Trans. ASME, 1987.

# Robot Welding Systems

Prof. Dr.-Ing. P. Drews - Dr.-Ing. G. Starke

During a period of more than twenty years robots nowadays have become one of the most important components in automated manufacturing processes.

First, most of robot installations had been assigned to increase the degree of automation in mass production particularly in the car industry. Here manual production procedures had been transferred to robots. Therefore resistance spot welding and material handling became the domains while arc welding with robots started very slowly due to the special requirements of torch motion and path accuracy.

Then, due to the continued improvements of robot technology in general and particularly in path control in the 1980's an explosive prevalence of robotic arc welding is to recognize.
Spot welding and arc welding became now the absolute domains of robot applications worldwide.

Through the years the number of robot installations was increasing rapidly. In 1988 we estimate a worldwide utilization of 324.350 robots. Contrary of the continuously increasing number of robots installed in production in the last years growth rates of robot installations decline worldwide. 37 % growth rate in 1987 changed to 26 % in 1988.

Nearly 70 % of the worldwide robot installations are contentrated in the five leading industrial nations Japan, US, Germany, Italy and France (Fig. 1).

In all those countries big car industries are located. Therefore we have to assume that the car industry itself and the subsuppliers are still the major users of robots today.

For example, in FRG about 35 % of all robots installed are working in car production in 1988. Particularly in spot welding applications but also in arc welding, material handling, assembling and coating.

According to Fig. 2 with nearly 40 % welding still represents the domain of robot utilization in FRG. This is similar to other countries. While spot welding with robots will decline to a certin saturation particularly in the car industry robotic arc welding becomes of increasing importance.

Today most of the arc welding robots are still installed in mass production. But due to the special need of high flexible manufacturing systems combined with market requirements for economical production and varied ranges the arc welding robot now also enters small and medium-sized companies. They are forced to find new approaches for

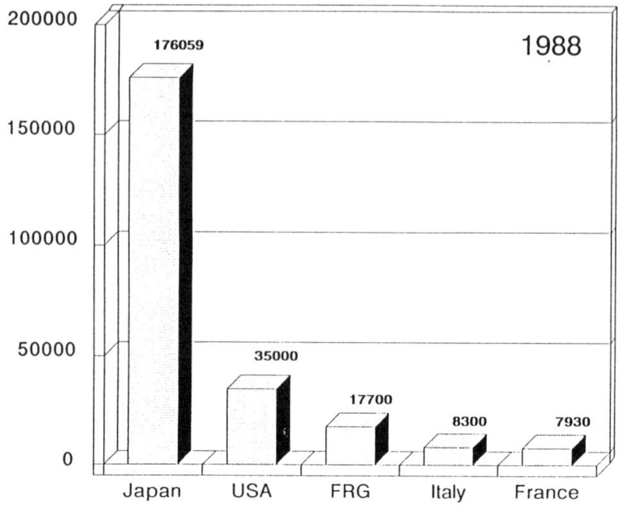


Fig. 1 World leading robot nations (IFR 1989)

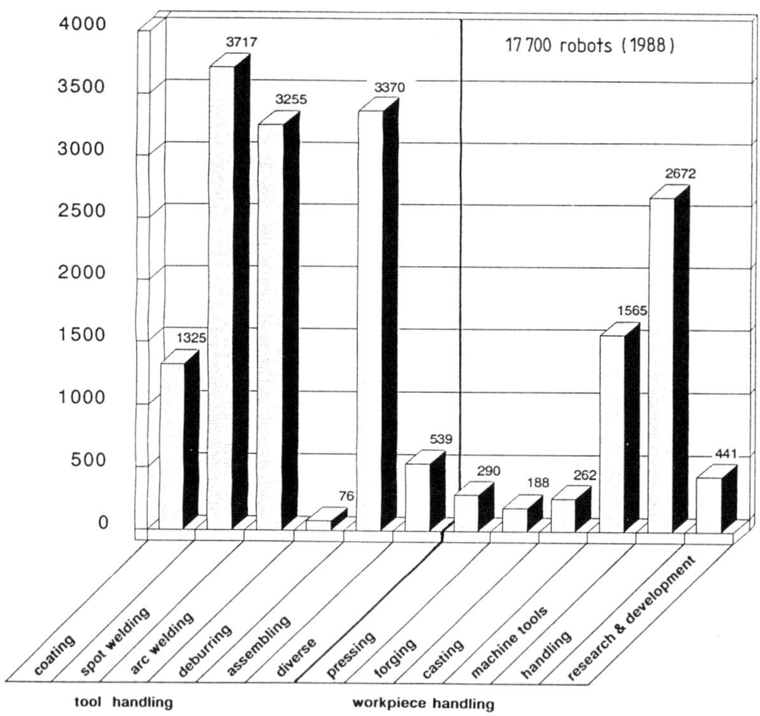


Fig. 2 Robot utilization in FRG (IPA 1989)

production automation in order to keep competitive.

A decision to introduce robotic arc welding in production has to be taken with special respect to technical and economical requirements particularly as far as small and medium-sized companies are concerned. The following study points out essential features.

An analysis of production costs of a typical workpiece out of a small batch production in a medium sized company represents very high expenses especially for welding (Fig. 3). So, it's obvious to reduce these costs first by special measures of rationalization. It was decided to replace manual welding by use of welding robots to decrease the production costs per piece and to improve productivity and weld quality.

The investment of a standard robot welding station today requires a minimum lot sizes to produce economically with a certain profit. The break-even-point between manual welding costs and robot welding costs was found for a minimum lot sized of 71 pieces at two shift operation (Fig. 4). An increase of the lot size further results in declining production costs per piece since a so called "optimal lot size" will be achieved.

Production costs of a standard robot station for welding are essentially influenced by the set up-time including fixturing, tooling, programming and testing.
Particularly programming with standard methods today takes a lot of time compared to the burning time of the arc.
The ratio is about 20:1 for standard welds.

Therefore, new approaches in programming arc welding robots are required.
A special contribution to solve this problem is the use of macros as a programming tool developed on our institute.
Macros are wellknown in computer techniques. In welding they represent a task-orientated sequence of torch motions including orientation combined with required process data to fulfil a defined weld. For instance welding of a corner or perhaps of a stiffener etc. (Fig. 5).

Macros for basic welding tasks can be preprogrammed and stored in a "Macro data base".
With free access to this data base programming becomes easier and saves time and money.

In our study by use of macro programming the break-even-point of production costs between manual and robot welding changed over to a minimum lot size of 3 pieces. A lot size of 12 pieces provides already reduced costs of nearly 34 % (Fig. 6).
The results give an idea of the very close connection of robot technology and economical aspects which has to be considered at each installation of a robot welding station particularly in small series production.

**COST OF A WELDED PART (%)**

| | |
|---|---|
| MATERIAL | 34 % |
| PREPARATION | 14 % |
| ASSEMBLING | 13 % |
| WELDING | 32 % ← |
| HEATTREAT. + PAINT. | 7 % |
| | 100 % |

Fig. 3 Cost structure

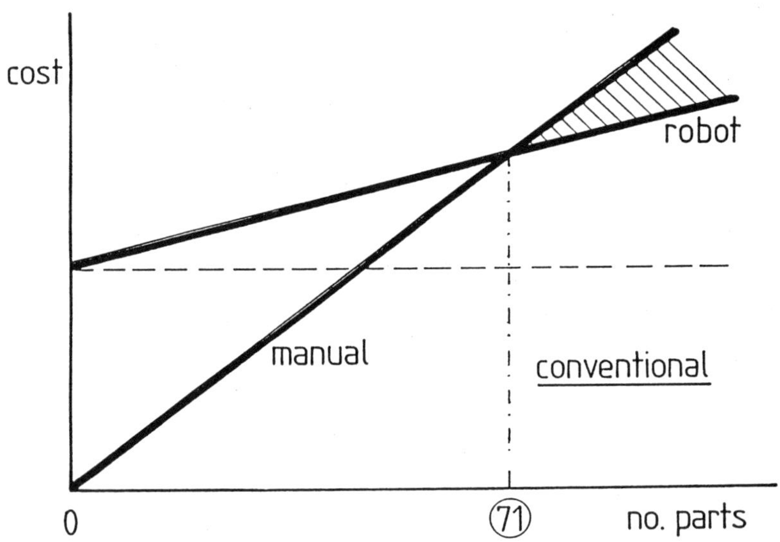


Fig. 4 Production costs for manual and robot welding

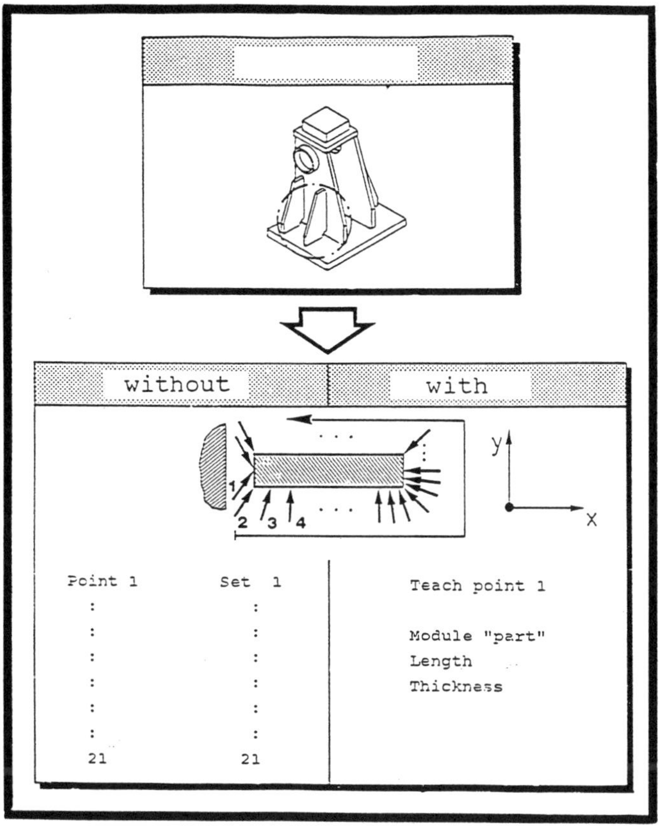


Fig. 5 Macro programming

Further, the study indicates a certain change in robotic welding engineering. Today the industrial user - especially in small and medium-sized companies - needs flexible, special adapted engineering solutions to assure high quality welds and to fulfil economical demands. Turnkey system with adapted technical solutions have replaced the arc welding robot as stand-alone system and the importance of the peripherical devices in a robot welding station becomes of more and more significance.
In order to minimize preparation and set-up time of a station continously and to assure quality of the products to be welded at the same time further steps in advanced robotic engineering are now the integration of intelligent sensor systems into robot systems and the use of CAD/CAM data already available in many industrial companies to improve programming.

According to those requirements as a first approach a modular control system for sensor guided robotic arc welding had been developed in the institute (Fig. 7). It is using CAD-data from the workpiece, its geometry and demands on the joint area. Those are taken into account as prior information for the welding process. Workpiece tolerances or insufficient joint preparation which do of course not correspond to the CAD data are compensated online by an integrated laser-sensor.

The modular structure of the control system represents a supervising controller called "co-ordinator" as central unit controlling and co-ordinating all components involved in the system like sensor, robot, welding equipment and CAD/CAM data base.
The robot is only responsible for torch motion along a preprogrammed welding path at a defined velocity and orientation. To adapt the torch movement to the actual path contour by sensor information it is necessary to provide communication between robot system, sensor and co-ordinator unit via powerful interfaces. External detected and calculated corrections of the TCP-movement by the sensor running ahead of the torch have to be taken into account for path calculation in the robot controller itself. Due to the delay between sensor measurement and torch positioning the actual velocity and tool position have to be available on request at the interface for calculation of the actual joint position data.

Those are then transformed into robot coordinates and transfered to the robot controller for path correction.

Determination and storage of the welding data are carried out by a special computer module as an interior part of the co-ordinator unit. Based on knowledge of the workpiece to be welded it determines the required weld start data with special respect to essential parameters like material, shielding gas, filler material, welding position, shape of the groove etc.

During welding the computer module continuously controls current, voltage and velocity depending on type and geometry of the joints.

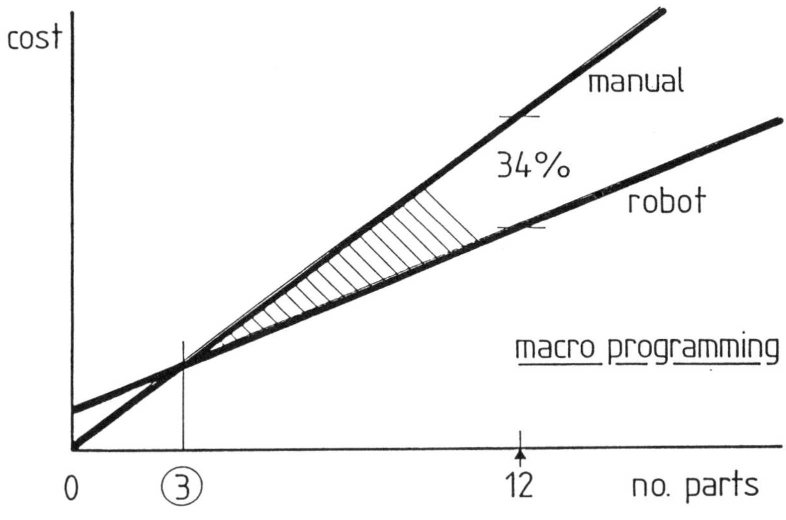


Fig. 6 Lowering of production costs with macro programming

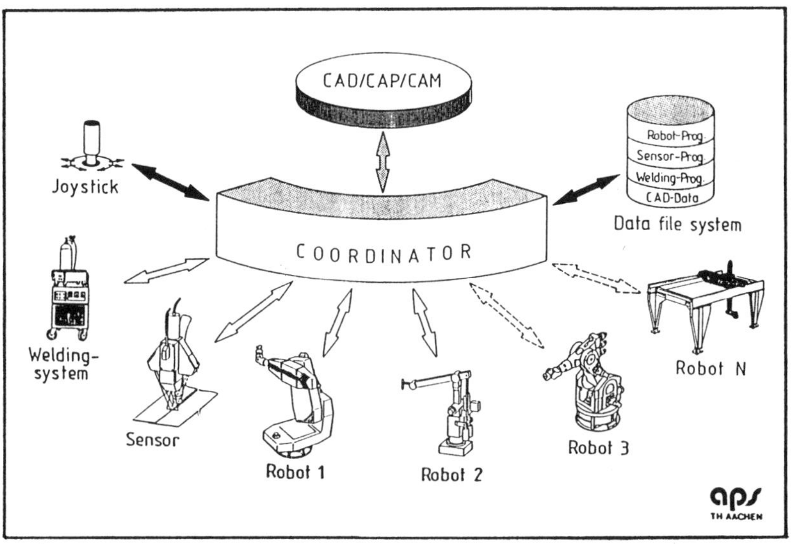


Fig. 7 Structure of a modular, sensor-guided control system for robotic arc welding with CAD/CAM-data input

Furthermore, predefined tolerance limits for relevant welding parameters are supervised in order to guarantee reproductable high quality welds.

In case of destortion the welding computer module generates error messages for the user. It stops the process if the predefinded limits are exceeded.

Starting a weld special operating conditions for the robot, sensor and welding equipment are prepared by the co-ordinator unit in consideration with CAD/CAM data and user inputs. They are transferred to the components involved in the system to start coordinated control of the entire system.

Besides the online-position data-calculation of the actual welding path by evaluation of the sensor signals the coordinator unit continuously supervised the information flow and status of all systems components. It initiates appropriate activities if necessary.

In order to assure high flexible applications of the control system special attention was given to a modular structure and the special design of the interfaces required for information and data exchange.

Due to a lack of an industrial interface standard all interfaces are custom-made to connect individual components. The chosen system design and the process computer capacity even permits the integration and co-ordination of several robots working in one flexible manufacturing cell.

## Conclusion

The industrial robot market and particularly robotic arc welding is a growth market. Continued improvements of this technology have enabled the robot to be developed to such a high degree of reliability that it can be seen today as a very important link in the production chains of the welding industry.

Nevertheless, the evolution of robot technology has to be continued with special respect to the users' requirements but also in view of "Advanced robotics".
Some of the major objectives for future R&D projects in advanced robotics can be suggested, as follows

- advanced manipulators
- improved control systems
- integration of transputer technology
- improved man-machine-communication
- offline programming with CAD/CAM data input and simulation

- advanced sensors
- integration of knowledge based systems
- intelligent interfaces
- FMS technology
- CIM integration.
- mobile robots.

References:


1. G. Spur: Robotertechnik in fortgeschrittenen Produktionssystemen, DVS-Verlag Düsseldorf No. 118, (1989), S.

2. D. Fabrizi: Die Roboterwelt, Robotermarkt 1990, MI-Verlag Landsberg (1990), S.10-12

3. M. Scweizer: Maßvoller Roboterzuwachs, ROBOTER (March 89), MI-Verlag Landsberg, S.30-34

# SENSING AND ADAPTIVE CONTROL OF ARC WELDING

T.ARAYA Mechanical Engineering Research Laboratory, Hitachi, Ltd.
Tsuchiura-shi, 300, JAPAN

Abstract

The first section of this paper discusses trends in arc welding automation, the sensors that are used in welding robots and automatic welders, the results from a questionnaire about seam tracking of weld lines, and adaptive control of welding conditions using those sensors, and future trends in this area. It will then talk about the two major sensor categories: contact and non-contact, and make a more detailed examination of seven sensors in those categories. It will then give a number of technical examples of seam tracking and adaptive control as well as some data on recent topical technologies.

Key words: Sensor, Sensing, Robot, Automatic welding, Adaptive control, Arc welding, Questionnaire, System, Seam tracking

## 1. Introduction

The electronics industry has grown rapidly in recent years and machinery, control, electrical and materials technologies have become highly sophisticated. Sensors, controllers, control software, robots and automatic machines are constantly being invented through an integration of these technologies, a development cycle is growing shorter and more sophisticated. That trend will continue unabated and with the use of computerization with welding robots, we will see true factory automation in the future. For that reason it is inevitable that there will be further progress made in research and development on sensors and adaptive control of those sensors.

The history of research and development on welding sensors and adaptive control is long. As far back as 1963 reports appear showing how welding temperatures are being controlled through the use of data from infrared thermometers or the use of adaptive control to hold the weld bead width constant[1]. Many different kinds of sensors and control technologies have been developed since, then but none of them satisfy all purposes, each has to be selected for the purpose it well be applied to.

Present reports that stand as good references are the "Development and Application of Sensors and Sensor Systems for Arc Welding in Japan",[2] submitted by the Japan welding Society's Welding Process Committee to the IIW XII Committee in 1982. As part II of that report, partial results of a 1989 survey are to be presented to IIW in Montreal in a report titled "Sensors and Control Systems in Arc Welding - The State of the Art in Japan."[3]

These reports systematically deal with the actual situation and future tasks in applying the sensors, seam tracking and adaptive control developed in Japan.

## 2. Trends in automatic arc welding

The national census conducted every five years says that there were 460,930 welders in Japan in 1975 and 335,600 in 1985, a decrease of 125,330, or 27.2 percent, in 10 years. This shows how the change in welding from manual to semi-automatic to robot-controlled requires a decrease in the amount of welding work. The results of a survey[1,2] on trends in the use of automatic and semi-automatic equipment, and number of welders and welding robots in Japanese shipbuilding and in the construction of ocean structures are shown in Figure 1.

From this we can see and increase in automatic and semi-automatic equipment since 1980, a rate that by 1988 had reached 80 percent of the total. The use of welding robots began to increase in 1982 and by 1988,80 were in operation. A survey by the Japan Industrial Robot Association[8,9] on the number of arc welding robots shows that there were 6,145 in 1988, as in Figure 2. That number probably reached 7,000 in 1989. The uses to which these robots were put in the 1988 fiscal year were 44.3 percent for car production, 10.8 percent for the production of metal products, 7.9 percent for general machinery, 6.7 percent for electrical equipment, and 4.1 percent for constructio n.They are now being widely used in production of all kinds.

The use of welding materials such as covered electrodes has been cut in half over a ten year period, but the increases in the use of solid wire and flux-cored wire are significant, this is shown in Figure 3. Since the development of metal wire in 1986, industrial use of flux-cored wire is more and more prevalent for building heavy machinery, steel girder frameworks, steel bridges, as well as ships.

Another area that cannot be overlooked as to automation and computerization of welding is that of electronics. A survey by the Japan Electronics Equipment Industry Association shows that continuous growth of 5-6 percent per year in the country's electronics production brought it to the 2.2 trillion yen level in fiscal 1989, a trend that is not forecast to change. The same trend is seen in the production of semiconductors, where the average unit price is going down every year so that what was 16 yen in 1983 will have reached 12 yen by 1988.

These surveys all show a rapidly increasing trend in arc welding

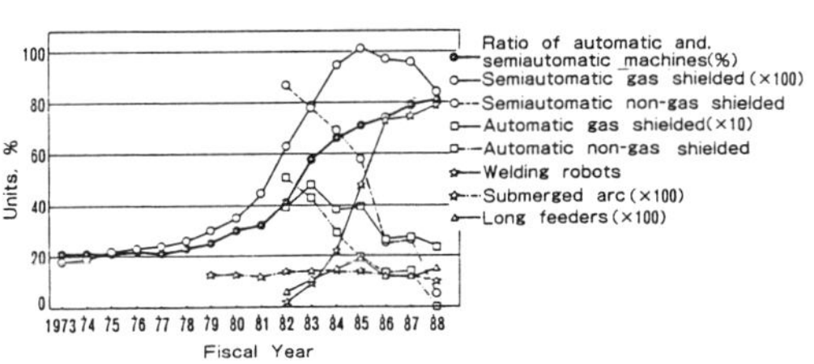


Fig. 1 Trends in the Ratio of Automatic to Semiautomatic and in the Number of Welding Machines

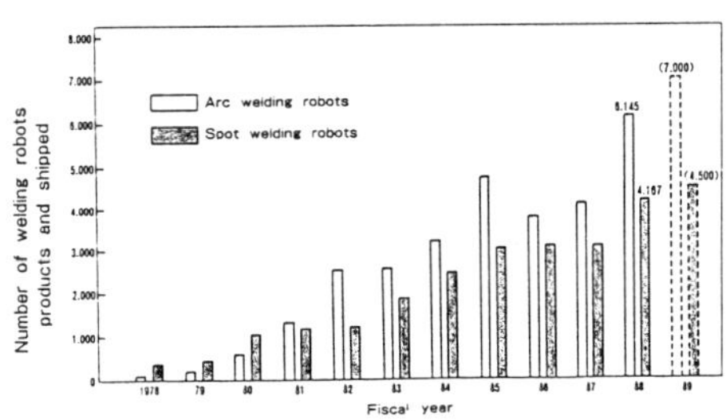


Fig. 2 Annual Trends in the Quantity of Welding Robots Produced and Shipped

automation over the past several years, a trend that will continue into the future.

Systems for automating MAG, TIG and submerged arc welders are shown in Figure 4. They consist of a robot, welding equipment a sensor to detect conditions, and a controller that organically connects and controls all these devices. Many kinds of control technologies are used in the controllers including N/C, AI, fuzzy and expert.

This report will examine automated systems, the sensors, the controller that uses those sensors to seam track weld lines, the technology for adaptive control welding conditions, and the applications of those systems.

## 3. Sensors for arc welding

The range of sensor applications is extremely broad, but as Table 1 shows, the sensors discussed here are those that detect welding materials and characteristics in order to control seam tracking of weld lines and adaptively control welding conditions.[2,3]

The following will be not considered in our discussion of sensors here.

1) Sensors for power supplies, wire feeders and other equipment used with welding machines.
2) Sensors for welding robots, manipulators and other automatic welding machinery.
3) Sensors for weld recording, weld inspection and other weld quality control applications.

Classifications of sensors that will be examined here, according to elements in their configuration, are listed in Table 2. They are categorized, basically, in two main groups, those of contact and non-contact types. Any developed in the past that are not used now have been excluded.

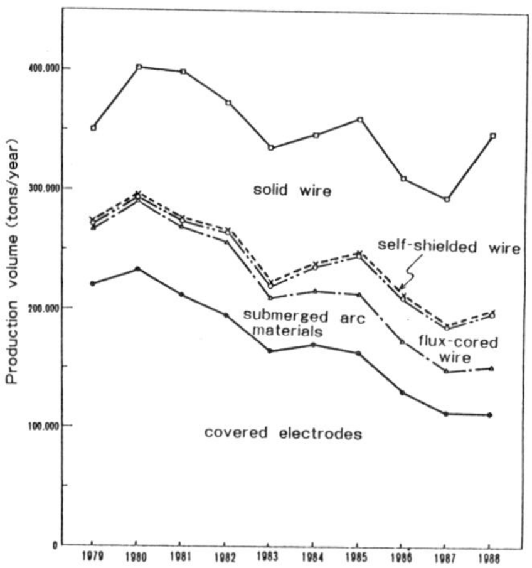


Fig. 3 Trends in Welding Materials Production

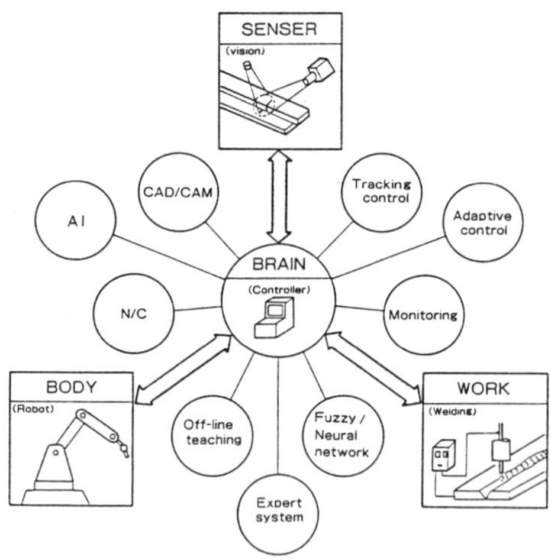


Fig. 4 Configuration of an Automated Arc Welding System

Table 1 Object Range of Sensors Used in Arc Welding

| | Detection object | Specific Example |
|---|---|---|
| Inside range | Welding materials | Contact or non-contact type sensors that can recognize welding position, groove shape, and obstructions. |
| Inside range | Welding characteristics | Sensors that can recognize arc length, wire extension length, arc shape, weld pool dimensions, external appearance of bead, penetration conditions, and arc sound. |
| Outside range | Welders | Sensors for detecting quantity of shield gas flow (pressure), coolant water pressure, current overload, and wire feed torque. |
| Outside range | Automatic welders | Sensors that control position by encoders or potentiometers and/or sensors that control speed by tachometers and accelerometers. |
| Outside range | Welding quality control | Sensors for inspecting the results of welding by X ray and ultra-sound and sensors for recording welding parameters. |

Table 2 Classifications of sensor

| | Sensor type | Units in the sensor configurations |
|---|---|---|
| Contact | Contact probes | Micro-switches, potentiometers, and differential transformers(DTF). |
| Contact | Electrode contact | Voltage and current for contact detection that is applied to the W electrode or the welding wire. |
| Non-contact | Thermal | Thermocouples, thermistors, photo-thermometers infrared thermometers. |
| Non-contact | Arc characteristics | Welding current, arc voltage, wire feed speed, number of shorts, number of peak current anomalies. |
| Non-contact | Electromagnetic | Electromagnetic eddy current detectors, magnetic detector by Hall devices. |
| Non-contact | Optics | Point sensors (phototransistors and photodiodes), linear sensors (CCD, MOS and PSD) and area (image) sensors (CCD, MOS, PSD and ITV). |
| Non-contact | Sound | Variable sound pressure or ultrasonic sound pressure detector probes. |

Changes in the use of sensors in Japan over the past 30 years are shown in Figure 5, in which point sensors equipped with single of multiple photodiodes or phototransistors were used early on but have been gradually being replaced since 1970 by linear and area sensors that detect light or heat patterns.

One-degree-of-freedom (linear movement) contact sensors equipped with differential transformers and potentiometers have been in use for many years because they are simple and inexpensive to operate and maintain. Two-degrees-of-freedom sensors were developed around 1967 and currently are widely used.

It has been about 10 years since through-the-arc and wire-touch sensors were developed and many are now in use. These are the easiest to operate with welding robots because they are self-contained, requiring no add-on components.

The recent expansion in the range of work applications has created composite sensors which are combinations of single sensors having different capabilities.

## 4. Control systems

Figure 6 shows that systems for controlling sensor-equipped arc welders to detect weld lines and/or weld conditions are configured from processors and other equipment that use the sense data to 1) judge weld lines and track welding torches attached to robots or automated machines, and 2) judge weld conditions and adjust all welder conditions to their optimum values.

The sensors detect 1) work shape errors, 2) setting errors, 3) groove shape change, 4) tack beads, 5) welding heat distortion and 6) jig error.

Factors disturbing sensors used in welding are 1) arc light,2) arc heat, 3) spatter, 4) electromagnetism, 5) bent or deflected electrode wires, 6) changes in wire speed, 7) worn electrode chips, and 8) changes in welding conditions or arc shape. Sensors must be either protected from these disturbances or equipped with functions to eliminate them.

Sensors should be: 1) adequately precise to handle welding processes, 2) unaffected by disturbance to the welding process (heat, light spatter, etc.), 3) sufficiently durable 4) inexpensive, 5) easy maintain, 6) general purpose. Welding control systems that use sensor data may be categorized according to the following purposes.

(1) Control seam tracking of weld lines
(2) Adaptive control of weld conditions
(3) Seam tracking and adaptive control

Many control systems have been developed and put into operation that use the sensors listed in Table 2, and the major ones will now be discussed.

### 4.1 Control seam tracking of weld lines

These paragraphs discuss mainly tracking control systems that use (1) electrode contact (wire-touch) sensors, (2) through-the-arc sensors, and (3) optical (area and image-processing) sensors mounted in welding robots.

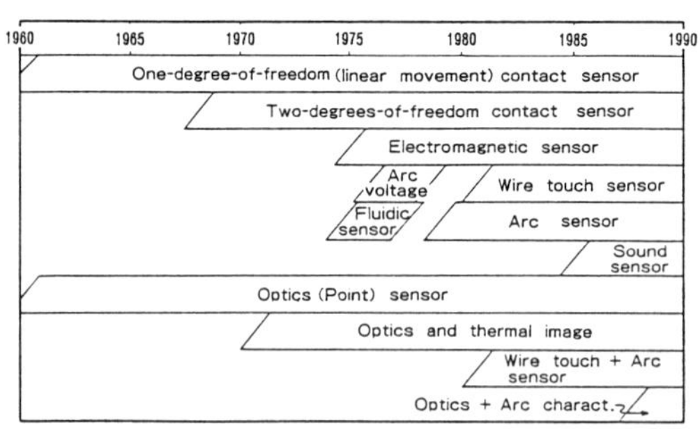


Fig. 5 Trends in Sensors Used in Arc Welding in Japan

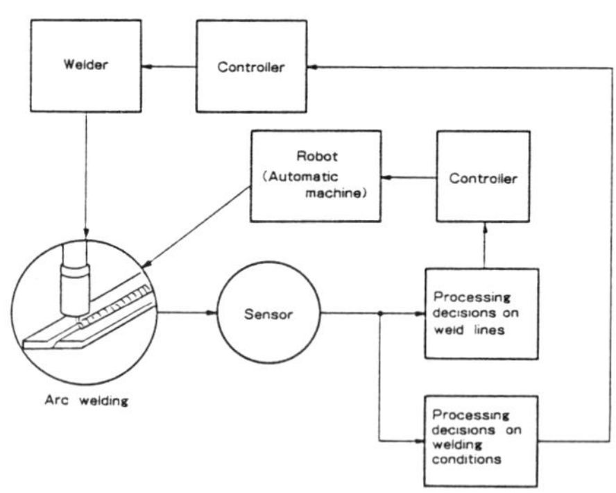


Fig. 6 Configuration of Sensor Control System for Arc Welding

Table 3 Specifications of Wire Touch Sensors

| Item | Specification |
|---|---|
| Surface condition of work and wire | surface resistance: 10kΩ or less |
| Detection speed | Approach speed: 1-50mm/s<br>Withdrawal speed: speed instructed to robot |
| Position calibration function | Parallel shift in multiple directions<br>Rotating shift in three dimensions |

### 4.1.1 Electrode contact sensors

These sensors detect changes in voltage or microcurrent (10mA or less), as Figure 7 shows, that flows when the welding wire comes in contact with the base metal and determines the coordinates of the contact point from the coordinate of the robot or automatic machine. They are usually called wire-touch sensors, wire-ground sensors, or touch sensors.

To obtain contact, voltage of 300-600V, 50 or 600Hz is applied. Touch sensor specifications are listed in Table 3 as an example.[21] A detector control function, which minimizes overshoot after the wire touches, gives a maximum detection speed of 50mm/s. It can also detect the position at which welding starts on the base metal, use its parallel shift function to teach weld line data to the robot, detect the weld start and stop points, and use its three-dimensional rotating shift function to adjust the welding line.

Because these sensors cannot be used during welding, a touch sensor detects the welding start point and in many cases causes a tracking composite method using arc sensors, discussed below, to seam track the weld line during welding.

### 4.1.2 Trough-the-arc sensors

These are generally known as arc sensors, ACC sensors, AVC sensors or through-the-arc sensors. When the welding torch is oscillated in the groove, the length of the wire extension changes and causes the welding current to change. This is depicted in Figure 8.

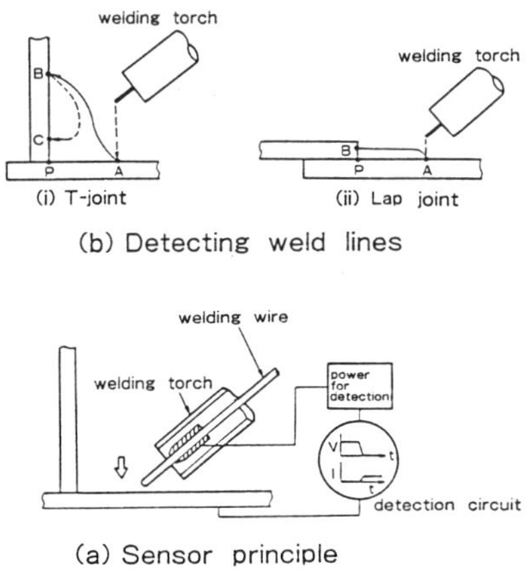

Fig. 7 Principles of Electrode Contact Sensors

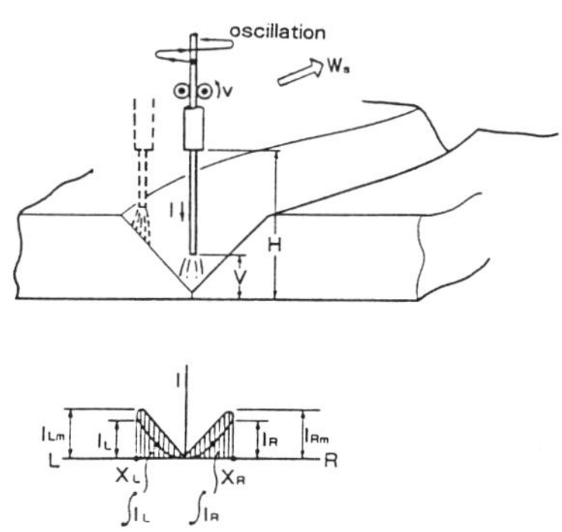

Fig. 8 Principles of Weld Line Tracking by Arc Sensor

Many tracking control systems have been developed that use the basic arc sensor concept, and some are shown in Figure 9. In many instances, the torch is oscillated by moving the machine, but the torch may also be held in place and the arc oscillated by electromagnetic force, or a gas will be blown on the arc and the thermal pinch effect causes the oscillations.[4] The tracking that is also shown in the figure is done in many different ways, too.

Methods of controlling seam tracking of weld lines process the data from the detection of the torch's oscillation movement to left and right to determine how far the torch is from the center of the groove and correct it. However, because the data that is detected is very unstable, new control principles are being developed that compare averages of data sampled from left and right, or compare data from near both the right and left edges, as shown in Figure 9.

There are some that detect welding current (I), arc voltage (V) and wire feed speed (v) to calculate the distance between the chip and the base H = f(I,V,v), and use the distance H rather than the current I, as shown in Figure 8, to control seam tracking of weld lines. Fuzzy control has also been introduced to these control types.[20]

Another system that controls the seam tracking of weld lines is shown in Figure 10. This system detects the welding current's frequency power spectrum.[15,16] The figure shows the position of the welding torch in relation to the welding line and the relation between welding current power spectra at that time. Figure 10 (b)

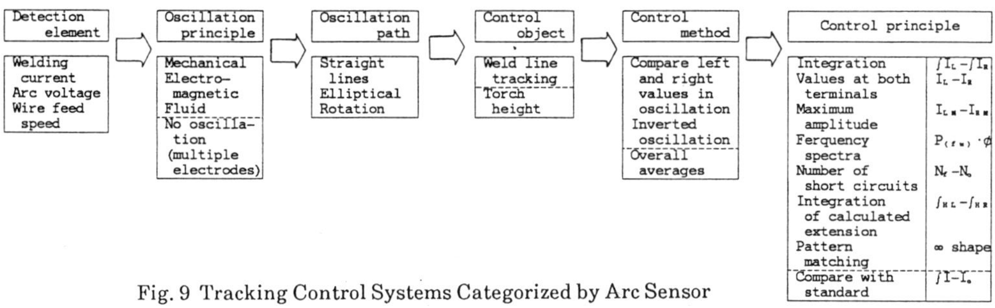


Fig. 9 Tracking Control Systems Categorized by Arc Sensor

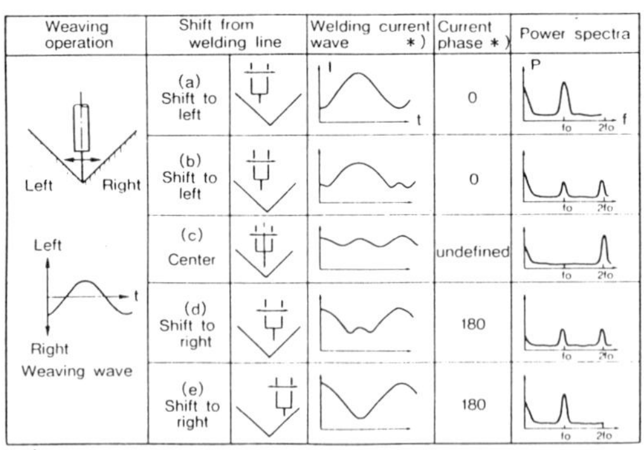


*) Current waves and phase angles for weaving wave

Fig. 10 Power Spectra of Welding Current According to Shift from Welding Line

shows that when the welding torch oscillates at the welding line center, frequency 2fo, twice as large as weaving, detects most of the welding current's power spectra, and that weaving frequency fo is almost unrecognizable. However, Figure 10 (a) and (c) show that when there is weaving away from the weld line, the weaving frequency's power spectra greatly increases and is determined to be either to the left or right from the current phase at that time.

MAG and $CO_2$ welding are based on this principle and can track lap-joint welded lines at a minimum thickness of 2mm.

4.1.3 Optical sensors

Many control systems have been developed that use optical sensors. Their results are systematically presented in Figure 11. They use the 2-pass method in which the sensor initially tracks the welding lines during the sensing period before welding, then welds by playing back the stored data. However, with the recent progress made in countering external disturbances that accompany welding, detection and welding may be performed simultaneously.

Different sensors are used in different ways; some representative methods are given in Figure 12.

(1) Tracking control by point sensor

Figure 12 (a) shows that the point sensor is oscillated or rotated so that it moves laterally across the weld line. The quantity of light received and the sensor position (coordinates) at that time are placed in two dimensions. The gap and groove position are detected and tracking is controlled.

(2) Tracking control by linear sensor

Figure 12 (b) shows that this is a method in which a semiconductor laser or other point light source irradiates the base, a linear sensor detects the light positioned at fixed angle $\theta$, and the emitted and received light are unitized. It uses the principle in which changes in distance $\Delta H$ from the base changes the linear sensor's light receiving position by the quantity $\Delta h$.

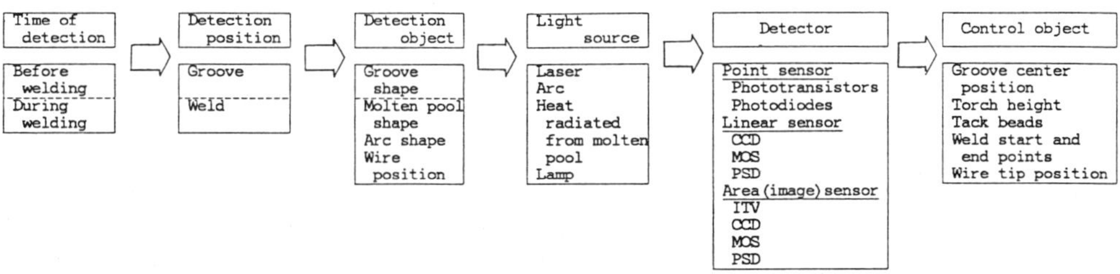


Fig. 11 Tracking Control Systems Categorized by Optics Sensor

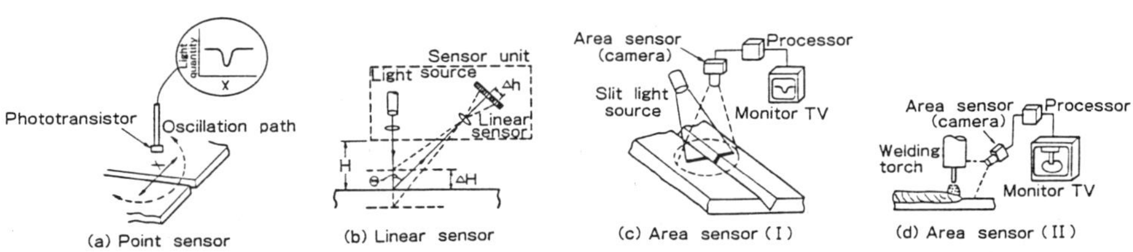


Fig. 12 Tracking Control by Optics Sensor

These sensors are used in either a method that controls welding torch height to fixed levels or one that oscillates or rotates the detection unit to detect the amount of change in base height (lap joint, V groove etc.) to control seam tracking if weld lines.

(3) Tracking control by area sensor

A method that receives the slit light irradiated on the base plate, and tracks the welding lines by the images processed is shown in Figure 12 (c). Another tracking method is shown in Figure 12 (d), in which an image is processed in the vicinity of the arc weld to determine weld position, wire position and molten pool shape. Each method uses many different techniques to prevent the disturbances that accompany welding.

The following are several examples of the image processing of slit light. The welding robot system is constructed from a welding robot, welding power supply, robot controller, welding torch and a unified sensor head and image processor.[17]

A sensor head structure that can rotate around the torch is shown in Figure 13. Two units of GaAIAs near-infrared semiconductor laser, an aspheric lens, and a cylindrical lens project the flat slit light. A Peltier device for cooling, and a thermistor for monitoring temperature work in combination to prevent the semiconductor laser's wave length from fluctuating.

An MOS imaging device is used for the light sectioned image in which a combination of intervention filters passes the semiconductor laser's oscillating wave lengths. Purge gas flows to prevent spatter and fumes from adhering to the windows in three places.

Image processing processes data from five picture elements and eliminates image changes caused by arc light and spatter. Mounted in an arc welding robot, it has weld line correction cycle of 0.5 seconds.

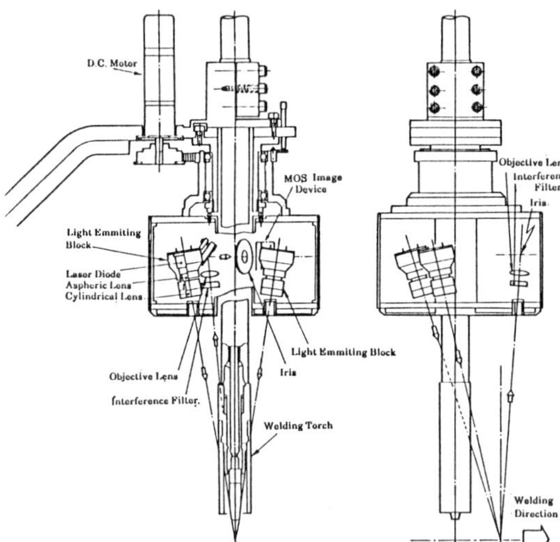


Fig. 13 Construction of Sensor Head

The field test shows that the image was sill clear after 100 hours of MAG welding at a current of 200-300A, and welded without an error in the lap joints of 300mm curvature and 1mm thickness.

4.2 Adaptive and tracking control

There is a lot of research and development being done in the areas of processing sensor data and controlling weld conditions to obtain optimum welding results. Combinations of adaptive control systems reported so far are shown in Figure 14. They are basically divisible into these components: (1) sensor (2) detection elements, (3) control elements, and (4) objects controlled. Many sensors use are characteristics, optics and thermal, and many of them control welding current and/or welding speed to optimize bead shape (width, height, penetration) and/or back bead shape.

4.2.1 Adaptive control by arc characteristics sensor

There are four types of detection elements for the arc characteristics sensors shown in Figure 14.

Methods that use the arc sensor to control the quantity of deposition are shown below. The method shown in Figure 15 calculates the deposition cross section area A using detections of groove width obtained when controlling the extension to keep it constant, oscillating the welding torch inside the groove, and reversing the oscillation direction when the torch is at the prescribed height. It then controls by detecting the amount of change $\Delta A$ in cross sectional area when the groove width changes, and calculating speed $v$ that gives a fixed bead height.[4] This control method detects changes in the welding deposition surface and controls the amount of filler wire fed out in the same way as TIG welding does.[4]

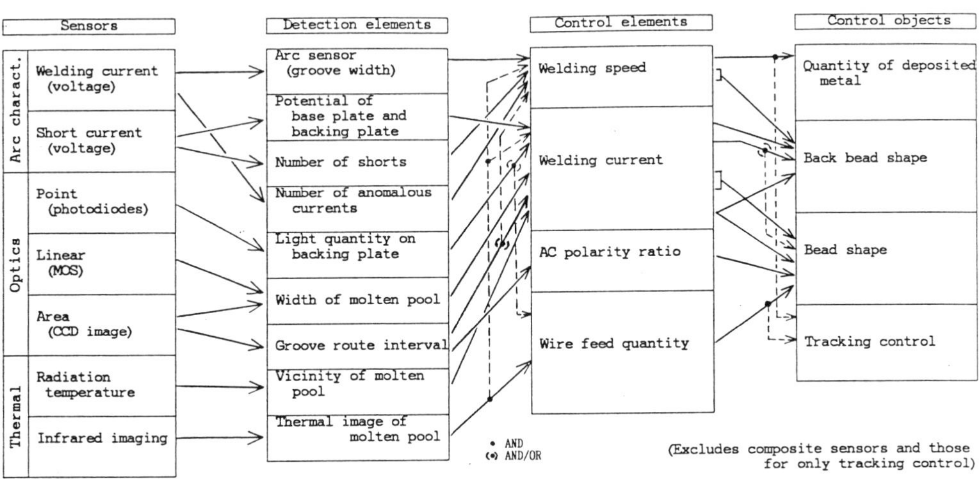


Fig. 14 Sensors for Adaptive Control Systems and Combinations of Control Elements

Additional methods are those that (1) control weld speed and keep the height of the back bead constant because the number of short circuit transfers in the root pass in relation to the back bead shape, (2) detect distortion in the weld current waveform (count numbers of anomalous peak current), control weld speed and keep back bead shape constant, and (3) detect the potential of the base plate and copper backing strip in one side SAW, control the welding current and keep back bead width constant.

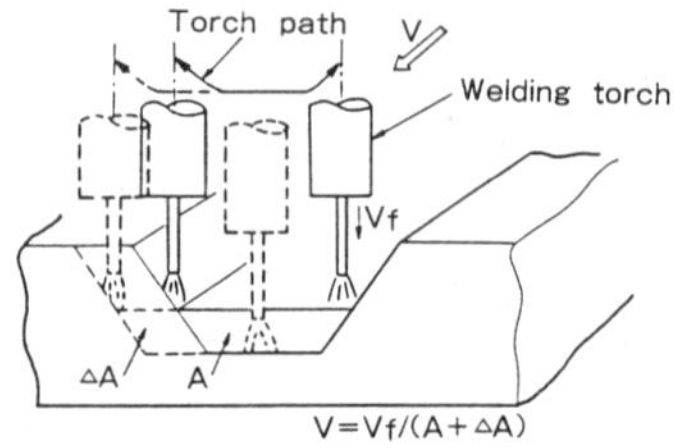

Fig. 15 Method of Controlling Quantity of Deposited Metal

4.2.2 Adaptive control by optical sensor

There are many reported types of adaptive control using optical sensors, but all have three types of detection elements (those that detect quantity of light on the back plate, those that detect width of the molten pool and those that detect the groove's route interval) as shown in Figure 14.

Methods that use point sensors to control back bead width and weld bead height have four CdS sensors. These sensors detect the quantity of SAW arc light passing through the back of the base plate, keep the quantity of light at appropriate levels by controlling the welding current, and maintain the back bead at a constant width. They, still more, control the amount of wire adding at a corresponding level with the welding current, and to maintaining the weld bead at a constant height.[4]

There are also adaptive control methods that use linear (MOS) sensors to detect the width of the molten pool and control the welding current to keep the bead shape constant.

Adaptive control methods that use area (CCD imaging) sensors to process images are those that 1) detect groove width from the image, and control welding current and speed to keep penetration depth and bead height (deposition quantity) constant, 2) detect groove width from the image and control the polarity ratio in AC pulsed MIG welding to keep the weld bead width and back bead width constant, and 3) detect the width of the molten pool from the image and control welding current to keep the penetration depth constant.

4.2.3 Adaptive control by thermal sensors

An adaptive control systems that detects temperatures in and around the molten welding pool is shown in Figure 14.

(1) Radiated temperature

The pulsed TIG method of welding stainless steel tubing is an all-position welding system that provides uniform back bead widths. A pulsed TIG welding device is configured from a welding head, infrared detector, temperature monitor controller and pulsed weld power supply. Infrared light radiating from the molten weld pool passes through a 2-3$\mu$m infrared filter and is made incident to a PdS photocondutor cell.

An He-Ne laser is used to adjust the thermal detector position and lens focal point. A 2mm thick SUS304 tube with a 50mm outside diameter is used to control the peak pulse current so that the assigned temperature remains constant, resulting in, as Figure 16 shows, a uniform weld temperature and constant back bead width W .

Note that excellent results can also be obtained with butt welds that have a 0-0.5mm gap and a root pass within the stainless steel tube's groove.

(2) Thermal imaging

Systems are being developed that make a thermal image on the molten pool to control seam tracking of weld lines and give uniform bead shapes. The units are configured as shown in Figure 12(d) from an infrared camera, image processor, welder and welder controller.

The seam tracking control of welding lines and welding heat input control methods perform binary-code processing (cut-out temperature T=t) of the detected thermal image, as shown in Figure 17. When the welding torch position changes, the thermal image warps, as Figure 17 (b) shows, but when the quantity of heat input changes, the thermal image's size changes as Figure 17 (a) shows. Figure 18 shows that this effect is used to adjust the welding torch position so that the thermal image's graphics center Xg and horizontal center Xc are the same. The standard thermal image area So, when welding is performed with appropriate heat input, is compare
d
with S, the detected binary-coded image area, to adjust welding speed v, welding current I and wire feed speed W, as Figure 18 shows. The results of tests on SUS304 using the TIG welding method, showed that the appropriate binary-coded processing temperature T was 1400℃ which is close to the solidification temperature. Experiments show the results of tracking control in which the 10mm wide grooved is moved away from the direction of carriage travel and also the results of welding on a base plate which changes in thickness. The control interval at this time was 0.8 seconds, tracking precision was 0.35mm and a welding bead, of almost uniform width, was obtained, as shown in Figure 19.

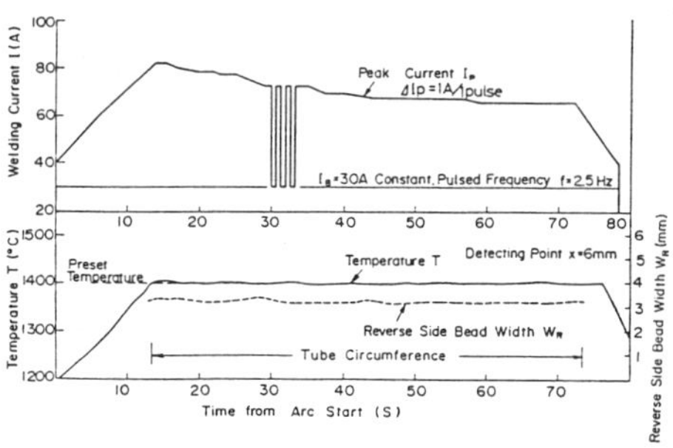


Fig. 16 Welding Current, Temperature and Reverse SideBead Width for SUS Small Tube Using Temperature Control

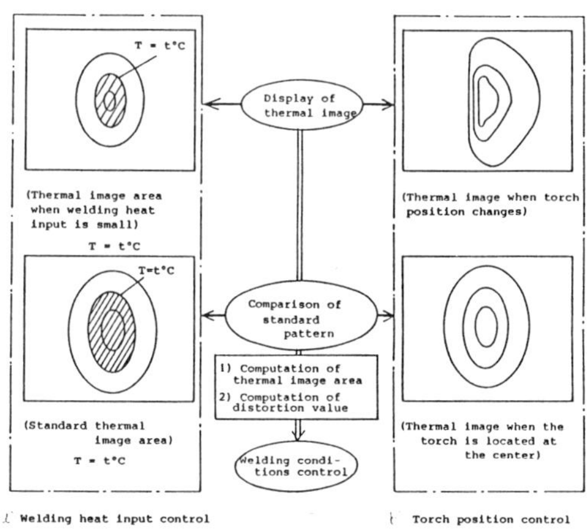


Fig. 17 Principle of Welding Control

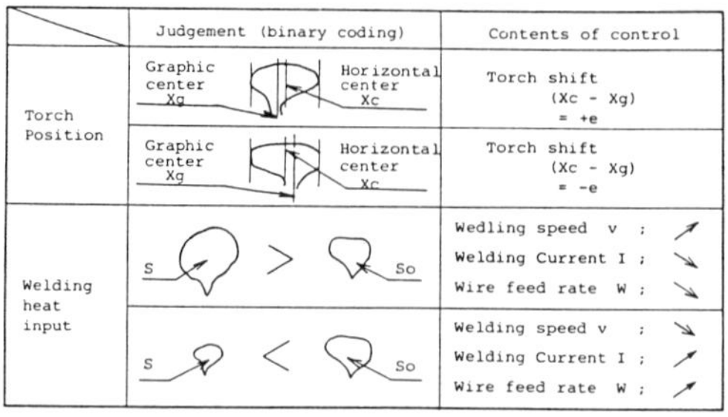


Fig. 18 Methods for Welding Control

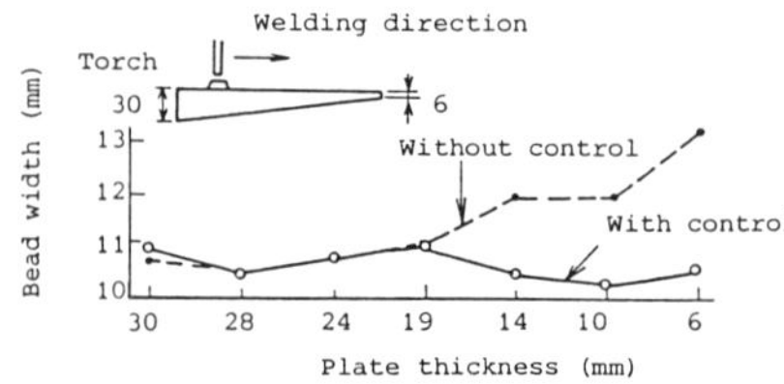


Fig. 19 Typical Welding Control

## 5. Future Tasks

Many types of sensor, tracking control and adaptive control technologies are now being developed. Future research and development will bring improvements in general purpose use and practicality of these technologies.

A survey by the Japan Welding Institute on the situation of arc welding robot use in 1987 showed that 73 percent of those robots were equipped with sensors, but as Figure 20 shows, only about half of them are performing to satisfaction.[10] A survey by the Japan Welding Society's Welding Process Committee on sensors used in arc welding during fiscal 1981 and 1989 shows problems with sensors in terms of accuracy, repeatability, spatter protection, durability, size, shape, and cost.[2,3] Seventy to eighty percent of these sensors are used for tracking control; there is still little use of them for adaptive control.

The Japanese welding environment, as Table 4 shows, will change greatly in the future in terms of its shortage of trained welder personnel, reduction in work hours, growth in the electronics industry and in internationalization. Progress in welding will require greater use of robots and computerization.

That concludes this discussion of the technologies for sensors, tracking control and adaptive control in Japan.

I hope it will be of some aid to welding progress.

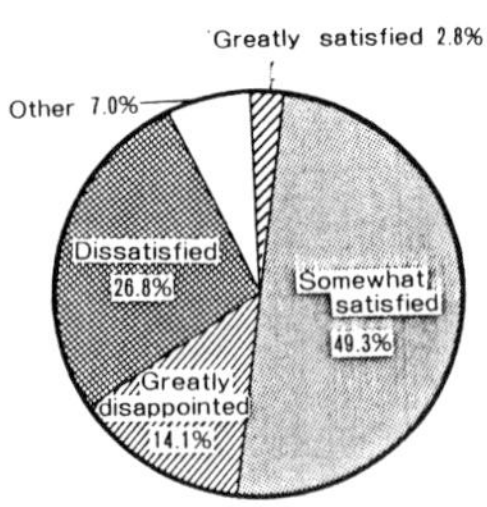


Fig. 20 Degree of Satisfaction with Sensors

**Table 4 Changes in Japanese Welding Environment**

| | Trend | Affects on Welding | Developments in welding materials |
|---|---|---|---|
| Social environment | ·Internationalization<br>·Individualization and diversification<br>·High ratio of value added<br>·Aging society<br>·Growth in the tertiary industries sector | ·Alternating flow<br>·Production of smaller amounts of products in greater varieties<br>·Higher quality and more advanced function<br>·Shortage of trained welders<br>·Reduction in work hours | ·Robot systems<br>·Mounting sensors<br>·Off-line teaching by personal computer<br>·Multiple welding systems<br>·Communication networks<br>·Diversification of heat sources<br>·Automation of non-destructive testing |
| Economic env. | ·Increase in value of yen, decrease in value of dollar<br>·Competition with manufacturers overseas<br>·Diversification of materials | ·Rationalization of production process<br>·Increase in overseas investment<br>·Diversification of processing techniques | |

References

1) Ramsey,P.W., Chyle,J.J., et al: Welding Journal,August,1963,337s~346s
2) Masumoto,I., Araya, T.,et al: IIW XII-C-031-82,1982
3) Nomura,H., Matsunawa,A.,et al: IIW XII- -90,1990
4) Nomura,H., Sugitani,Y.:Joytech, April, 1989,59-65
5) Murakami,E., Hisagai,K.: Welding Technique, Vol.38,No.1(1990),120-126
6) Inoue,K.:Journal of The Japan Welding Society,Vol.51,No.9(1982),735-741
7) Questionnaire, Welding Technique, Vol.35,No.12(1987),76-82
8) Report of survey, Robot, No.70(1989),32-43
9) Report of Survey, Welding Technique, Vol.37,No.10(1989),108-110
10) Report of Survey, Welding Technique, Vol.37,No.5(1989),92-100
11) Sakai,Y., Suga,T.:Welding Technique, Vol.37,No.11(1989),117-122
12) Yasuoka,T.: Welding Technique, Vol.37,No.11(1989),78-82
13) Yamamoto, Y.: Journal of The Japan Welding Society, Vol.58,No.4 (1989),272-279
14) Nishida,Y., Furuya,K.,et al : Welding Technique,Vol.37,No.11 (1989). 123-127
15) Araya, T., Tsuji,M., et al: Hitachi Review, Vol.34,No.1(1985),27-32
16) Nakajima,J.,Araya,T., et al: IIW XII-954-86,1986
17) Hirai,A., Shibata,N., et al : Advances in Flexible Automation and Robotics-Vol II,1055-1060,1987
18) Kozono,Y., Onuma,A., et al: Quarterly Journal of The Japan Welding Society,Vol.6,No.1(1988),110-115
19) Takahashi,Y., Koide,H.,et al: IIW XII-998-86,1986
20) Fujimura, H., Ide, E.,et al : Quarterly Journal of The Japan Welding Society, Vol.5,No.3(1987),311-316
21) Manual of Hitachi Welding Robots, 1989

REAL TIME X-RAY INSPECTION

HEIDT, H.
Bundesanstalt für Materialforschung und -prüfung (BAM), Berlin

Abstract
Real time inspection of welds is an important consequence of automated manufaturing if the safety of the product has to be assured by process-independent quality control. Several equipment types have the potential for weld inspection, but a comparison is difficult because of non-existent specification standards. The operation of real-time systems is very different from the conventional X-ray film, so neither the common radiography standards nor the film-related Image Quality Indicators (IQI) can be applied with satisfactory results. A new set of standards must be developped by Standard Committees to provide reliable weld inspection with radioscopic systems. In the future, automated image interpretation and decision finding must be included in real-time inspection.
Key words: Real time inspection, radioscopy, equipment specification, standardization, image interpretation

1. Real time inspection - Why?

Industrial production has changed to automatic processes dramatically during the recent two decades. Tube welds and automotive bodies are manufactured without human assistance, and very highly stressed parts in aircraft and space industry are assembled under robotic control to maintain optimum reliability and flawless manufacturing.

But even with optimal process control, non-destructive testing (NDT) will not loose its value for safety and economy: For the optimization of the process and for the initial testing of prototypes, NDT is the first and economic method of choice. Additionally, many products with highest safety demands have to be inspected on a 100 % basis. Typical examples in welding industry are pipelines, nuclear plants, but even critical joints in bridges and vehicles.

For automated production, a conventional manual NDT procedure may be a costly and delaying experience. The key word is "Real time inspection". Concerning the field of X-ray inspection, the main obstacle to a faster inspection was the use of radiographic film, which needs exposure time, unpacking from cassettes, processing (manual or automatic) and visual interpretation. Under good standardized conditions, the whole procedure takes typically twenty minutes. The visual interpretation may cause unconstant, unreliable decision finding on behalf of the momentaneous condition of the human interpreter. Reliability investigations with several interpreters show - regarding on the type of object and the importance of the defect - a statistical reliability of 40 to 80 %. To say it clearly: No interpreter, whatever his or her qualification is, gives you a guarantee for a 100 % defect recognition rate. Additionally, the reliability of defect recognition may vary from day to day - or even from minute to minute.

This is where real time inspection may show its second advantage, if it is completed with automated image interpretation. The quality of interpretation is constant and predictable. A specified system will give you a defined defect sensitivity, but even a predictable ratio of false calls, if the selected sensitivity level is close to the limits of the Real time system.

## 2. Real time inspection - How?

In 1986, a Working Party of IIW Subcommission VA started some fundamental work in describing the state of the art in X-ray Real time inspection under the guidance of Dr. Ron Halmshaw, Great Britain. Several reports with definitions and typical data were published. For a general classification of these systems, three groups were formed:

- The *open screen type* (group 1) has a fluorescent screen which is projected to a TV camera (fig. 1). A light intensifier can be placed at the input of the camera. An advantage of group 1 systems is the unlimited size of the conversion screen

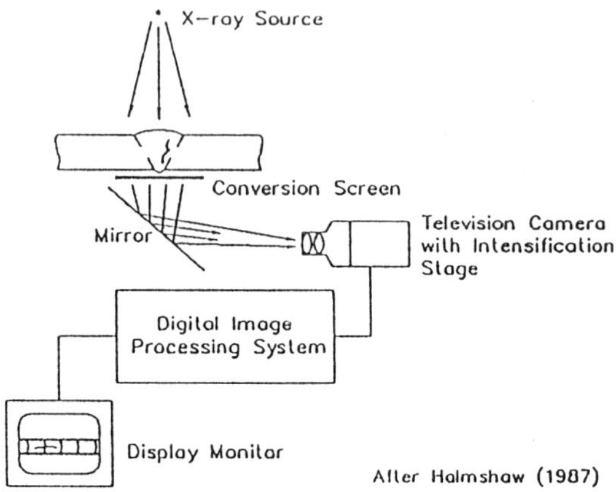


Fig. 1: Radioscopy system Group 1 - "Open" screen type system

- *X-ray image intensifier systems* (group 2) are today the mostly used type of radioscopy systems (fig. 2). The main advantage is the high light amplification of the intensifier tube which gives bright images under real time conditions.

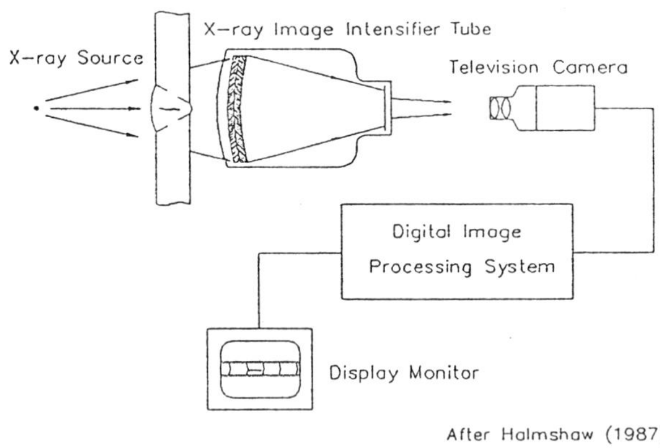


Fig. 2: Radioscopy system Group 2 - X-ray image intensifier system

- *Linear array systems* (group 3) are coming with the development of improved line sensors (fig. 3). They are inexpensive and give good contrast image because of the collimated beam geometry.

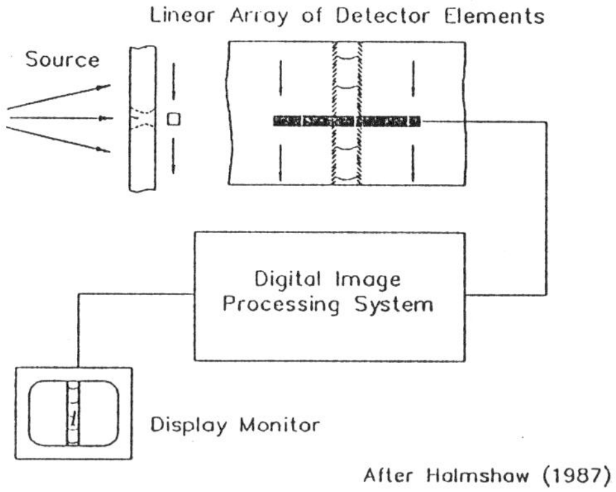


Fig. 3: Radioscopy system Group 3 - Linear-array sytem with Framestore

Especially the group 3 systems demonstrate that digital image processing is a typical feature of real time inspection. Even group 1 and 2 systems frequently use image enhancement to improve signal-to-noise ratio and to modify the contrast for better defect recognition.

Up to this point, there may be some confusion about the vocabulary used in real time inspection. *Radioscopy* is the real time alternative to film *radiography*: It gives a transmission X-ray projection of the specimen. The documentation can be on photographic film or on digital mass storage, i. e. optical discs. The old term *fluoroscopy* refers to the type of conversion screen which is fluorescent.

Real time inspection is not necessary bound to images. *Radiometric inspection* gives one or several line scans of the specimen. The density fluctuations can be interpreted as wall thickness variations or voids. For certain tasks in weld inspection, radiometric inspection can be a very fast and economic solution.

In some cases, welds are not accessible from both surfaces. A new and attractive solution to this problem may be Compton scatter imaging which is applicable from one surface and provides an image (fig. 4) of one (or several) distinct layers in depth, parallel to the surface of the specimen. Because of the low intensity of scattered radiation, this method becomes slow for distant layers.

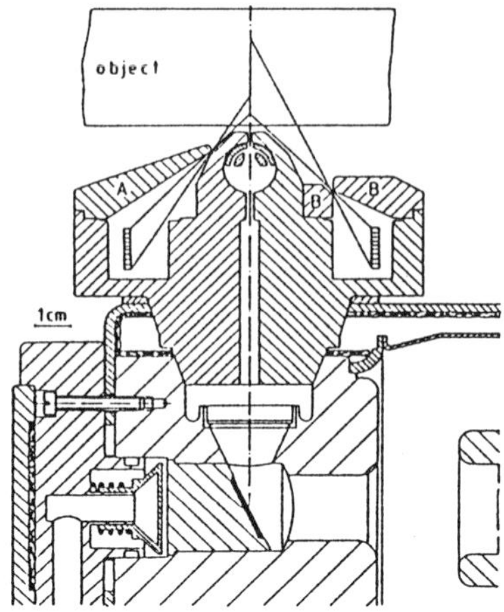


Fig. 4: Principle of the Comscan detector head (after PHILIPS)

The further considerations are concentrated on systems of group 1 to 3 which have a high potential for standard requirements.

A substantial comparison of equipment specifications is not possible because of divergent procedures for measurement. A future solution will be discussed in chapter 4.

## 3. Real time inspection - Why not?

Nowadays, the standards for radiographic weld inspection are generally based on the X-ray film as an imaging and documentation device. Even the control of radiographic image quality by penetrameters and wire type IQI's (Image quality indicators) is ideally suited to the control of film application and film processing.

It is impossible or unadequate to transfer the needs for film radiography without modification to real time inspection. Some main reasons are:

- The rules for minimum geometric distances of the set-up are based on a specific film property called inherent unsharpness. The equivalent value of radioscopic systems is significantly higher and gives reason to direct magnification with small focus tubes.
- The rules for maximum radiation energy for film radiography are based on physical contrast laws (specific contrast). For real time systems, the exposure time is limited to one TV frame period. This causes a high amount of noise in the image. The radiation energy must

be a compromise between noise and contrast, but both parameters can be modified in a wide range by electronic controls and image processing.
- With existing real time equipment, gamma ray sources like Iridium 192 and Cobalt 60 are hardly applicable because of the low intensity and relatively large size.

From these basic considerations it should be clear that the replacement of film radiography by real time inspection needs a good knowledge of system specifications and task-related investigations about the equivalence of defect detectability. The present standards for radioscopic systems (ASTM, BS etc.) have the character of guidelines for the user and give some confidence in the operation of systems, but - unlike the standards for film radiography - they are not a straight forward advice for the set-up of systems and the choice of parameters.

To sum up the state-of-the-art shortly:

- Radioscopic imaging *devices* perform substantially different from radiographic film
- Radioscopic *systems* can perform equal to radiographic film in many applications, if set-up and parameters are chosen optimally.
- At the moment, the optimization is not a straight-forward procedure because of *missing standards* and *uncertain specifications*.

## 4. Real time inspection - What to do?

In spite of the mentioned situation, real time inspection is used and shall be used wherever it is necessary and permitted. On the other hand it should be understood that the performance is not sufficiently described by IQI's and simple test specimen which were designed for film radiography. The use of electronics implies new possibilities but even new unvertainities, like electronic and quantum noise, limited dynamic range, unlinearities and ageing of components. These uncertainities are an urgent call for specification and control procedures which has to be answered by standardization committees. IIW SC VA has proposed a three part standard structure:

Main title: Qualification of radioscopic systems for weld inspection

Part 1: Quantitative measurement of imaging properties for radioscopic systems

Part 2: Qualitative control and stability of imaging devices

Part 3: Requirements for radioscopic systems for weld inspection

The three parts are directed to different readers: Part 1 describes the responsibilities of the equipment producer to specify his product correctly; part 2 shall be applied by the user of equipment to demonstrate the actual status of equipment compared to First-day-performance. For part 2 simple IQI and fast electronic checks shall be performed periodically.
Part 3 will give straight-forward rules for the planning and parameter setting of real-time systems. The specifications of parts 1 are used to determine the geometric set-up, the useful radiation energy and eventually necessary image enhancement. This part is directed to both the user and the producer of equipment. A system which operates according to part 3 shall be accepted for weld inspection without further discussion, as standardized weld radiography on film is accepted generally today.

## 5. Real time inspection - The future!

By now, the considerations were concentrated on radioscopic *imaging*. The electronic image may be evaluated by a human interpreter on a monitor screen, but automated inspection will necessarily include automated and immediate interpretation to take the necessary actions in the production process. Digital image processing has demonstrated the capability for automated interpretation, if a training set of weld defects is available. The processing is speeded up by concentrating to critical defects, which have to be defined before the learning phase (fig. 5). New processor technologies have shifted the price range of image processing to attractive hardware costs. This will result in an economic 100 % automated weld inspection wherever it is required. For the future, real time X-ray inspection will be a reliable and economic tool for quality assurance in process and product control.

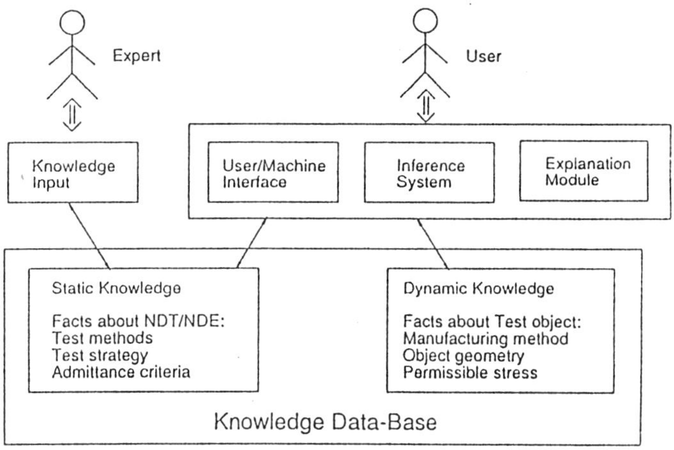


Fig. 5: Knowledge based automatic interpretation of weld defects

PART THREE

# Material Science Concepts in Joining

# THE OXYGEN AND NITROGEN ABSORPTION OF IRON WELD METAL DURING ARC WELDING

T. KUWANA
Department of Materials Processing, TOHOKU University

Abstract
Using thermodynamic data, the free energy changes for dissociation reactions of several gases are obtained at high temperature. The relationship between the partial pressures of several gases and temperature is shown under 0.1 MPa (1 atm) atmosphere. Oxygen and nitrogen absorptions into liquid iron in equilibrium state are discussed thermodynamically. The effects of the welding condition and the oxidizing gas partial pressure in the welding atmospheres of $Ar - O_2$ and $Ar - CO_2$ on the oxygen content of the purified iron weld metals during gas metal arc welding are showed. The effects of the welding condition and the nitrogen partial pressure in the welding atmospheres of several gas mixtures on the nitrogen content of the iron weld metals during gas tungsten arc welding and gas metal arc welding are showed. Using those results and thermodynamic data, the behavior of the oxygen and nitrogen absorptions into the iron weld metal during arc welding is discussed.
Key words: Arc welding, Gas-metal reaction, Oxygen, Nitrogen, Iron, Weld metal.

## 1. Introduction

Gases absorbed into steel weld metal during arc welding have an effect on the properties of the weld zone. The oxygen in steel weld metal causes nonmetallic inclusion and porosity, and has significant influences on the microstructure and the mechanical properties of weld metal. The nitrogen in steel weld metal causes porosity and affects the microstructure and the properties of weld metal, too. Recently, those gases absorptions have been investigated from view points of the contamination of welds by the arc atmosphere and the control of the microstructure of weld metals in a gas shielded arc welding process. Moreover, there have been several attempts to utilize the nitrogen in weld metals because the nitrogen stabilizes austenite structure of stainless steel.

Author has been studying on the absorptions of the gases into the various weld metals during arc welding for a long time. Especially, the effects of the welding conditions and the gases partial pressure in the welding atmosphere on the oxygen and nitrogen contents of the iron weld metals were systematically studied. At the same time, using thermodynamic data, the behavior of the gas absorption into the iron weld metal was discussed.

In this paper, the dissociations of gases under high temperature, solubilities of oxygen and nitrogen in liquid iron and the behavior of the oxygen and nitrogen absorptions into purified iron and mild steel weld metals under the welding atmospheres of several gas mixtures are shown. Using those results and the thermodynamic data obtained by equilibrium measurements, gas-metal reactions in the iron weld metals during arc welding are discussed.

## 2. Dissociation of Gas

The temperature of the gas in arc column rises over 5000 K under arc welding. Molecular gas ($O_2$, $H_2$ and $N_2$) dissociates into atomic gas (O, H and N) in arc column. The dissociation of a diatomic molecular gas ($X_2$) is shown by the equation (1).

$$X_2 = 2X \qquad (1)$$

The equilibrium constant (K) of the reaction is given as follows;

$$K = Px^2/Px_2 \qquad (2)$$

where

$Px_2$ : partial pressure of molecular gas ($X_2$)

$Px$ : partial pressure of atomic gas (X)

Entropy and enthalpy of formation of various gases at 298 K and their entropy and enthalpy data at 298 ~ 5000 K give the standard Gibbs free energy changes ($\Delta G°$) for dissociation reactions of gases.

The free energy change can be described approximately by the equation (3).

$$\Delta G° = \Delta H° - T\Delta S° \qquad (3)$$

where $\Delta H°$ and $\Delta S°$ are changes of enthalpy and entropy. $\Delta H°$ and $\Delta S°$ for dissociation reactions of several gases in Table 1 are given by calculating thermodynamic data.

Table 1. Enthalpy and entropy changes in various reactions for computation of free-energy change at 2000 ~ 5000 K.
$\Delta G° = \Delta H° - T\Delta S°$

| Reaction | $\Delta H°$ (J / mole) | $\Delta S°$ (J / deg·mole) |
|---|---|---|
| $H_2 = 2H$ | 460,410 | 122.93 |
| $O_2 = 2O$ | 513,460 | 134.89 |
| $N_2 = 2N$ | 966,670 | 136.44 |
| $CO_2 = CO + {}^1/_2O_2$ | 271,580 | 81.17 |
| $H_2O = H_2 + {}^1/_2O_2$ | 253,550 | 58.87 |
| $NO = {}^1/_2N_2 + {}^1/_2O_2$ | -89,370 | -12.30 |

The relationship between K and $\Delta G°$ is shown with the gas constant (R) by the equation (4).

$$\ln K = -\Delta G° / RT \qquad (4)$$

Using the equations (2) and (4) and data in Table 1, the relationship between the partial pressure of gases and temperature is obtained under 0.1 MPa(1 atm) atmosphere, as shown in Fig. 1.

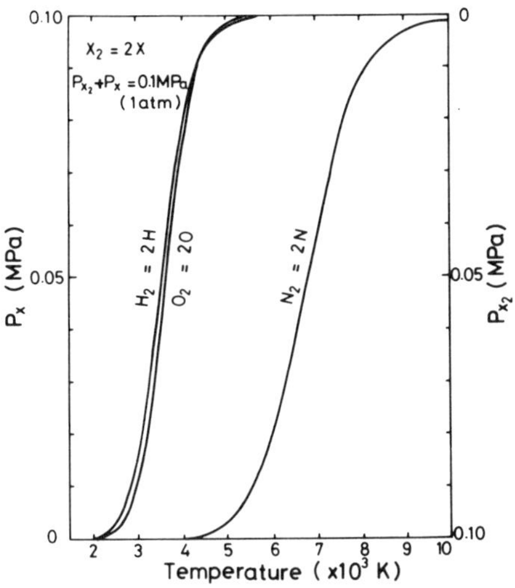


Fig. 1. Dissociations of oxygen, hydrogen and nitrogen.

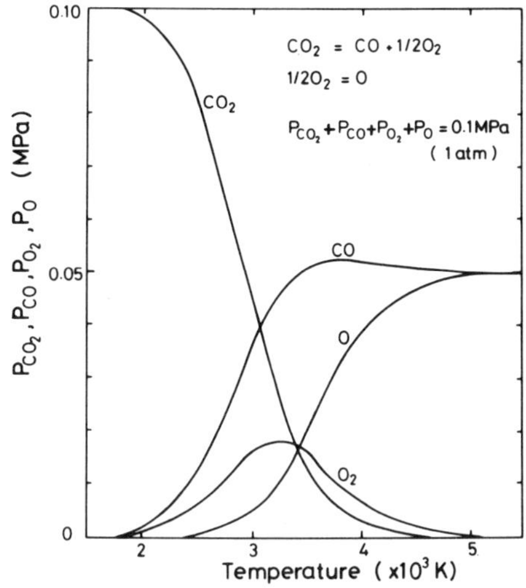


Fig. 2. Dissociation of carbon dioxide.

Molecular oxygen ($O_2$) and hydrogen ($H_2$) dissociate easily at temperatures above 3000 K. On the other hand, the dissociation of molecular nitrogen ($N_2$) is more difficult than oxygen and hydrogen.

The dissociations of those gases are very small at temperatures below 2000 K. Fig. 2 shows the dissociation of carbon dioxide.

## 3. Absorption of Gas into Liquid Iron under Equilibrium State

### 3.1 Solubility of oxygen in liquid iron

Molecular gas $O_2$ and atomic gas O coexist in oxygen atmosphere at a temperature as before, and the partial pressures of these gases ($Po_2$, Po) determine the oxygen content of liquid iron.
The relationship between oxygen gases in atmosphere and the oxygen in liquid iron is shown as follows;

$$^1/_2\ O_2 = O \qquad K_1 = Po\ /\ Po_2^{1/2} \qquad \cdots\cdots\cdots\cdots\cdots (5)$$
$$^1/_2\ O_2 = \underline{O} \qquad K_2 = [\%O]/\ Po_2^{1/2} \qquad \cdots\cdots\cdots\cdots\cdots (6)$$
$$O = \underline{O} \qquad K_3 = [\%O]/\ Po \qquad \cdots\cdots\cdots\cdots\cdots (7)$$

where

$Po_2$ : partial pressure of molecular oxygen
Po : partial pressure of atomic oxygen
$\underline{O}$ and [%O] : oxygen content in liquid iron

If the absorption reaction of oxygen into liquid iron is in equilibrium state, the equations (5), (6) and (7) should hold simultaneously. $\Delta G_2°$ and $K_2$ of the equation (6) were given by the equations (6.a) and (6.b) respectively.

$$\Delta G_2°(cal) = -27930 - 0.57T \qquad \cdots\cdots\cdots\cdots\cdots\cdots\cdots (6.a)$$
$$\log K_2\{=[\%O]/\ Po_2^{1/2}\} = 6105/T + 0.125 \qquad \cdots\cdots\cdots\cdots (6.b)$$

where the pressure of oxygen is indicated with the unit of atm.

The oxygen solubility in liquid iron is limited by the formation of liquid iron oxide phase. The solubility of oxygen in liquid iron was shown in Fig. 3 by Chipman et al. The solubility depends only upon the temperature and is given by the equations (8) and (8.a).

$$FeO(l) = \underline{Fe}(l) + \underline{O} \quad \cdots\cdots\cdots\cdots \quad (8)$$

$$\log K_4 \{ = [\%O] / a_{FeO} \} = \log [\%O] = -6320/T + 2.734 \quad \cdots\cdots \quad (8.a)$$

where

$a_{FeO}$ : activity of iron oxide in slag, $a_{FeO} \doteqdot 1$

By using the equations (8.a) and (6.b), it is known that the oxygen solubility is about 0.23 wt% and the oxygen pressure is about 1 mPa ($10^{-8}$ atm) at the temperature of 1873 K (1600 °C). A pressure is so low that the absorption behavior of oxygen into liquid iron can not be measured accurately by direct methods using oxygen gas.

## 3.2 Solubility of nitrogen in liquid iron

The relationship between nitrogen gases in atmosphere and the nitrogen in liquid iron is shown as follows;

$$^{1}/_{2}\ N_2 = N \qquad K_1 = P_N / P_{N_2}^{1/2} \quad \cdots\cdots \quad (9)$$

$$^{1}/_{2}\ N_2 = \underline{N} \qquad K_2 = [\%N] / P_{N_2}^{1/2} \quad \cdots\cdots \quad (10)$$

$$N = \underline{N} \qquad K_3 = [\%N] / P_N \quad \cdots\cdots \quad (11)$$

where

$P_{N_2}$ : partial pressure of molecular nitrogen

$P_N$ : partial pressure of atomic nitrogen

$\underline{N}$ and [%N] : nitrogen content in liquid iron

The equation (10) was measured by Pehlke and Elliott, as shown in Fig. 4.

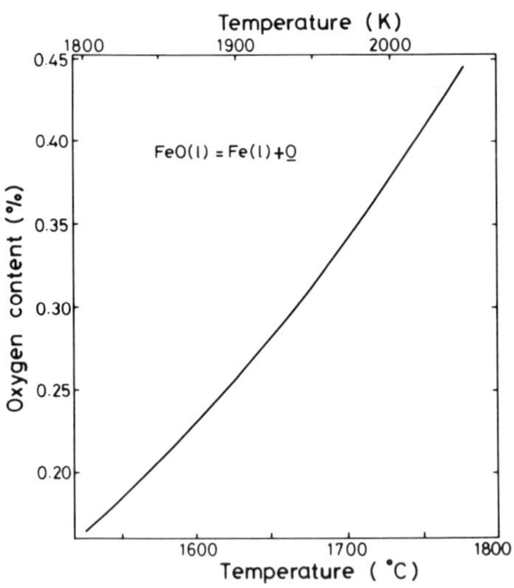


Fig. 3. Solubility of oxygen in liquid iron. (Chipman et al.)

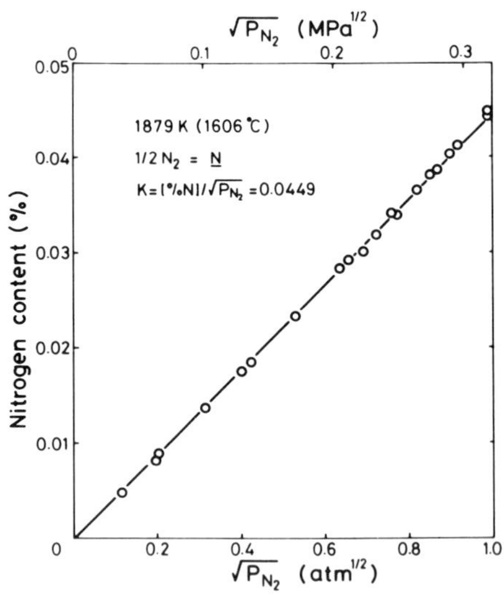


Fig. 4. Relationship between the nitrogen content of liquid iron and the square root of the nitrogen pressure. (Pehlke et al.)

The relationship between the square root of the partial pressure of the nitrogen (atm) and the nitrogen content is linear. The dissolution of the nitrogen into liquid iron obeys Sieverts' law. $K_2$ (atm) of the equation (10) was given by the equations (10.a).

$$\log K_2 = -188/T - 1.248 \qquad \cdots\cdots\cdots\cdots\cdots \quad (10.a)$$

## 4. Oxygen Absorption of Iron Weld Metal

Purified iron plate was welded using purified iron electrode wire of 1.6 mm diameter and an automatic welding machine under the $Ar-O_2$ and $Ar-CO_2$ welding atmospheres in a closed chamber.

### 4.1 Temperature of molten weld pool

A bead-on-plate weld was deposited on purified iron plate in the $Ar-O_2$ welding atmosphere. Temperature of molten weld pool was measured with thermocouple. The temperature of the molten iron weld metal during this gas metal arc welding is in the range from 1860 to 1960 K and does not seem to be concerned in the welding current in the range 150 ~ 350 A. The mean temperature is about 1910 K.

### 4.2 The effects of the welding conditions on the oxygen content of iron weld metal

Fig. 5 shows the relationship between the oxygen contents of the iron weld metals and the welding current in the welding atmospheres of $Ar-O_2$ ($P_{O_2} = 0.005$ MPa) and $Ar-CO_2$ ($P_{CO_2} = 0.003$ MPa) during gas metal arc welding.

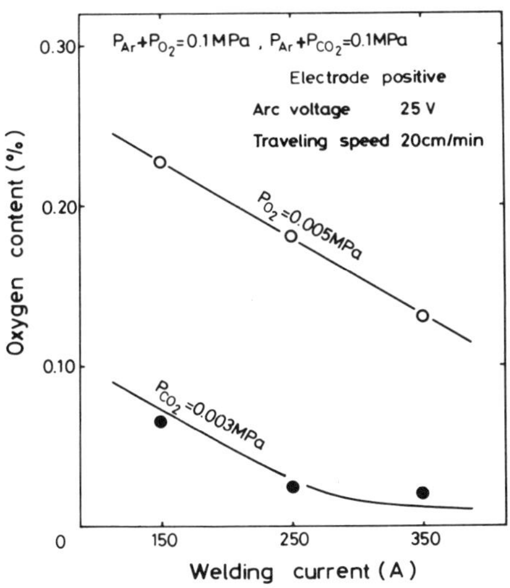


Fig. 5. Relationship between the oxygen content of the iron weld metal and welding current.

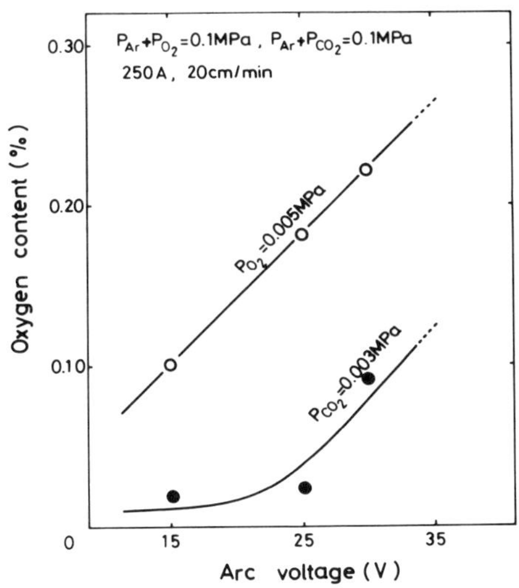


Fig. 6. Relationship between the oxygen content of the iron weld metal and arc voltage.

The oxygen contents of the weld metals decrease with increase of the welding current in both welding atmospheres. The reduction of the oxygen content may be attributed to the increase in transfer rates of the molten drops and the increase in the volume of molten weld metal with the welding current.

The effect of the arc voltage on the oxygen contents of the weld metals is shown in Fig. 6. The oxygen contents of the iron weld metals increase with an increasing arc voltage. With an increasing arc voltage, the arc length extends and the surface area of the molten weld pool becomes larger. Consequently, the reaction between the molten iron and the oxygen gas in the welding atmosphere may be promoted by an increasing arc voltage.

On the other hand, the oxygen contents of the weld metal decrease slightly with an increasing traveling speed.

### 4.3 The effect of oxidizing gas partial pressure in the welding atmosphere on the oxygen and carbon contents of iron weld metal

Fig. 7 shows the relationship between the oxygen contents of the iron weld metals and the partial pressure of oxidizing gases in the welding atmospheres during gas metal arc welding. In the $Ar-O_2$ welding atmosphere, the oxygen contents increase with a rising oxygen partial pressure up to about 0.010 MPa. Above 0.010 MPa, the oxygen contents are a constant value of about 0.26 wt%.

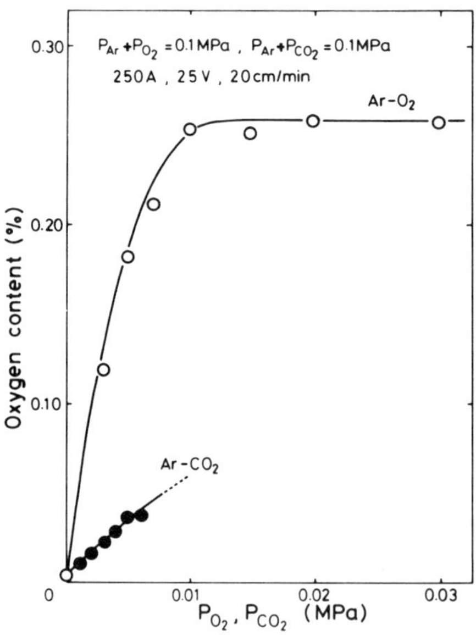


Fig. 7. Relationship between the oxygen content of the weld metal and oxygen partial pressure and carbon dioxide partial pressure.

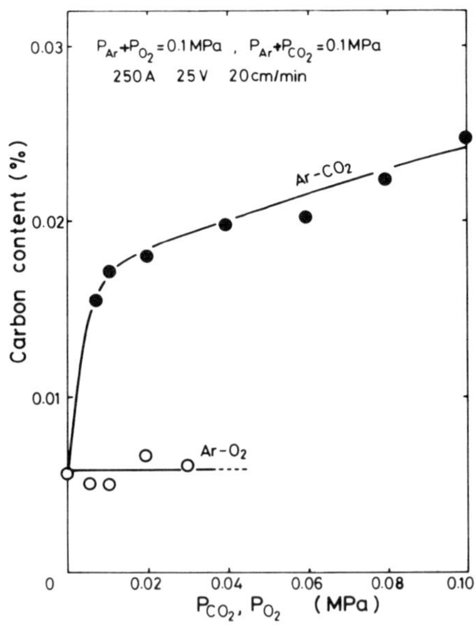


Fig. 8. Relationship between the carbon content of the iron weld metal and oxygen partial pressure and carbon dioxide partial pressure.

No porosity was observed in all weld metals welded in the $Ar-O_2$ welding atmosphere. In the $Ar-CO_2$ welding atmosphere, the oxygen contents gradually increase with a rising carbon dioxide partial pressure up to about 0.006 MPa. Above 0.006 MPa carbon dioxide partial pressure, a great deal of porosity was observed in the weld metal, and the analysis of the oxygen in weld metal was impossible. The oxygen contents of the iron weld metals in the $Ar-CO_2$ welding atmosphere are lower than those in the $Ar-O_2$ welding atmosphere.

Fig. 8 shows the relationship between the carbon contents of the iron weld metals and the partial pressures of oxygen and carbon dioxide in welding atmospheres. In the $Ar-CO_2$ welding atmosphere, the carbon contents of the weld metals increase considerably with arising carbon dioxide partial pressure up to about 0.02 MPa, and then gradually increase to a value of 0.024 wt% at 0.1 MPa(1 atm) carbon dioxide pressure.

It is clear that both oxygen and carbon in the iron weld metal were absorbed simultaneously from carbon dioxide in the $Ar-CO_2$ welding atmosphere. Therefore, the porosity in the iron weld metal is caused by CO reaction during arc welding.

As was stated previously, $O_2$ and $CO_2$ dissociate in an arc column according to following equations;

$$O_2 = 2O \quad \cdots\cdots\cdots\cdots\cdots\cdots \quad (5')$$

$$CO_2 = CO + {}^1/_2\, O_2(O) \quad \cdots\cdots\cdots\cdots\cdots\cdots \quad (12)$$

The oxygen and the carbon may be absorbed into molten weld metal from gases according to following equations;

$$O_2 \rightarrow 2\underline{O} \quad \cdots\cdots\cdots\cdots\cdots\cdots \quad (13)$$

$$O \rightarrow \underline{O} \quad \cdots\cdots\cdots\cdots\cdots\cdots \quad (14)$$

$$CO_2 \rightarrow CO + \underline{O} \quad \cdots\cdots\cdots\cdots\cdots\cdots \quad (15)$$

$$CO_2 \rightarrow \underline{C} + 2\underline{O} \quad \cdots\cdots\cdots\cdots\cdots\cdots \quad (16)$$

$$CO \rightarrow \underline{C} + \underline{O} \quad \cdots\cdots\cdots\cdots\cdots\cdots \quad (17)$$

where

$\underline{O}$ : oxygen in the weld metal

$\underline{C}$ : carbon in the weld metal

In the $Ar-O_2$ and $Ar-CO_2$ welding atmospheres, the difference of the oxygen contents in the iron weld metals may be related to the intensity of the oxidizing potential in those gases.

The oxygen absorbed into the molten iron weld metal separates from molten iron as nonmetallic inclusion at the solidification of the molten pool. It is supposed that the inclusions are mainly oxide with a composition of FeO by analyzing with an EPMA. According to the phase diagram for the Fe - O system, solid iron has very little solubility of oxygen. However, a lot of oxygen dissolves in liquid iron. In the equilibrium state, following equations were obtained;

$$FeO(l) = \underline{Fe}(l) + \underline{O} \quad \cdots\cdots\cdots\cdots\cdots\cdots \quad (8)$$

$$O_2 = 2\underline{O} \quad \cdots\cdots\cdots\cdots\cdots\cdots \quad (6')$$

As was stated previously, the oxygen content of liquid iron is about 0.23 wt% and the oxygen pressure is about 1 mPa ($10^{-8}$atm) at 1873 K (1600 °C) in the equilibrium state. Therefore, there is no possibility of porosity formation in the molten pool by gaseous oxygen.

The oxygen contents of the iron weld metals increased gradually with an increasing oxygen partial pressure up to about 0.01 MPa and then maintained a constant value of about 0.26 wt% in the Ar - $O_2$ welding atmosphere as shown Fig. 7, and the mean temperature of the molten iron pool was about 1910 K. If the reaction of oxygen absorption into the iron weld metal were in the equilibrium state, the oxygen contents of all iron weld metals would show a constant value of about 0.26 wt% corresponding to the temperature of 1910 K regardless of the oxygen partial pressure. However, the molten iron pool could not absorb oxygen enough up to the level of the solubility of liquid iron during welding at low oxygen partial pressure, while the molten iron pool could absorb oxygen enough from the welding atmosphere corresponding to the oxygen solubility at 1910 K at high oxygen partial pressure. During the solidification of the molten iron weld pool, the oxygen remaining in the weld metal combined with iron atom and formed the ferrous oxide inclusions.

## 5. Nitrogen Absorption of Iron Weld Metal

### 5.1 Nitrogen absorption of iron weld metal during gas tungsten arc welding

#### 5.1.1 The effects of the welding conditions on the nitrogen content of iron weld metal

Fig. 9 shows the relationship between the nitrogen contents of the purified iron weld metals and the welding current during gas tungsten arc welding. The nitrogen contents of the weld metals decrease with an increase of the welding current.

Though the nitrogen contents decrease slightly with an increase of the arc length, the degree of the decrease is very small. Since the arc voltage increases linearly with the arc length, the nitrogen content is considered to be almost independent of the arc voltage.

Blowholes and pits are detected in all those weld metals by observations of the bead surfaces and X-ray radiographic testings.

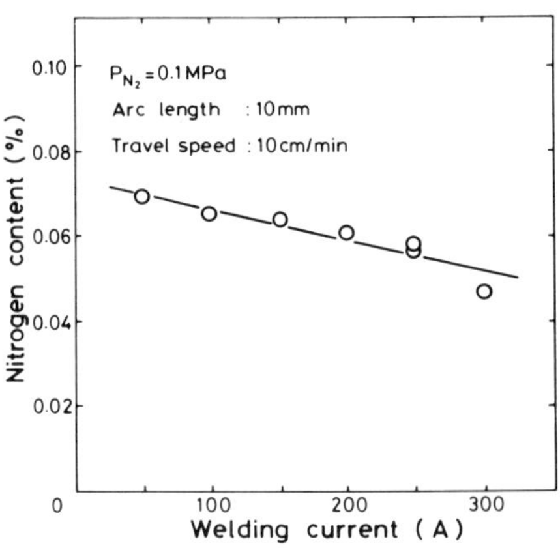


Fig. 9. Relationship between the nitrogen content of the iron weld metal and welding current.

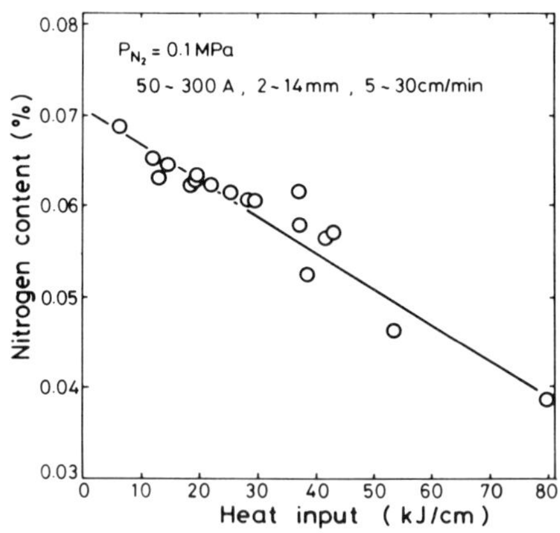


Fig. 10. Relationship between the nitrogen content of the iron weld metal and the welding heat input.

From the equations (10) and (10.a), the solubility of nitrogen in liquid iron at a nitrogen atmosphere of 0.1 MPa (1 atm) is 0.0445 wt%, 0.0448 wt% and 0.0458 wt% at 1809 K(the melting point of iron), 1873 K and 2073 K respectively. The temperature of the molten iron weld metal during this gas tungsten arc welding was measured to be about 1863 ~ 1933 K. The nitrogen contents of the weld metals are higher than the value of equilibrium solubility at the measured temperature of the molten weld metal and all of those weld metals have a lot of blowholes and pits. These phenomena suggests that a large amount of nitrogen was absorbed into the molten weld pool from arc atmosphere and then some amount of the absorbed nitrogen was evolved during cooling process.

Under a nitrogen welding atmosphere of 0.1 MPa, the nitrogen content of the weld metal seem to be controlled mainly by the evolution of nitrogen from the molten iron during cooling process.

Fig. 10 shows the relationship between the nitrogen contents of the weld metals and the heat input at a nitrogen welding atmosphere of 0.1 MPa. The nitrogen contents of the weld metals decrease with an increase of the welding heat input. A decrease of the cooling rate may prolong the period of time before solidification of the molten weld metal and may bring an increase of the amount of evolved nitrogen. This can explain the decrease of the nitrogen content with the increase of the welding current. This fact suggests that the cooling rate determines the nitrogen content of the iron weld metal obtained at a nitrogen welding atmosphere of 0.1 MPa.

### 5.1.2 The effect of the welding atmosphere on the nitrogen content of iron weld metal

In order to examine the effects of the welding atmospheres on the nitrogen content of the iron weld metal, the purified iron specimens were welded varying the nitrogen partial pressure ($P_{N_2}$) in the welding atmospheres of $N_2$, $N_2$ - Ar, $N_2$ - He and $N_2$ - $H_2$.

Fig. 11 summarizes the relationship between the nitrogen contents of the weld metals and the square root of the nitrogen partial pressure ($P_{N_2}^{1/2}$). In the nitrogen welding atmosphere ($P_{N_2} \leqq 0.1$ MPa ),the nitrogen content is anomalously high at low nitrogen pressure. The relationship between $P_{N_2}^{1/2}$ and the nitrogen contents in the $N_2$ - Ar and the $N_2$ - He welding atmospheres is almost same. The nitrogen content in the $N_2$ - $H_2$ welding atmosphere is lower than in other atmospheres. The relationships between $P_{N_2}^{1/2}$ and the nitrogen contents of the weld metals are not linear in all welding atmospheres. Therefore, the nitrogen absorption into the iron weld metal does not obey Sieverts' law (the square root law).

The nitrogen absorption of the iron weld metal during an arc welding process is characterized by the following three facts;

(a) The nitrogen content of the iron weld metal is considerably higher than the value of the nitrogen solubility in liquid iron except in the $N_2$ - $H_2$ welding atmosphere.

(b) The nitrogen content of the iron weld metal depends on the kind of the welding atmosphere even at a same nitrogen partial pressure.

(c) The nitrogen absorption into the iron weld metal does not obey Sieverts' law.

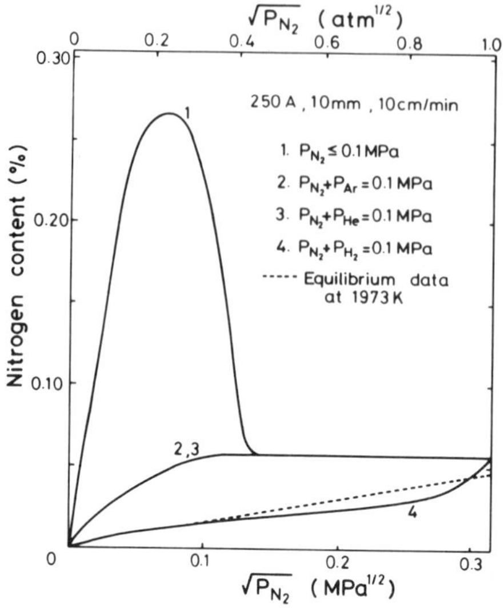


Fig. 11. Relationship between the nitrogen content of the iron weld metal and the square root of the nitrogen partial pressure during gas tungsten arc welding.

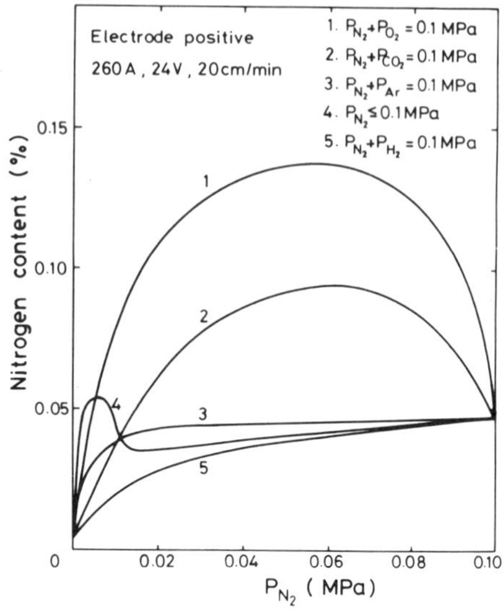


Fig. 12. Relationship between the nitrogen content of the mild steel weld metal and the nitrogen partial pressure during gas metal arc welding.

In equilibrium state, the partial pressures of gases($P_{N_2}$, $P_N$ ) determine the nitrogen content of liquid iron regardless of the kind of the atmosphere, and the nitrogen absorption obeys Sieverts' law.

In arc welding process, the equations (9) and (11) must be taken into consideration especially. Since the temperature of gaseous phase in arc column is considered to be 5000 ~ 7000 K, a large amount of atomic nitrogen (N) may be present in gaseous phase by reaction $1/2N_2 \rightarrow N$ as shown in Fig. 1. When the gases contact the surface of the molten weld pool of about 1900 K, they are cooled and the recombination of atomic nitrogen gas by reaction $N \rightarrow 1/2N_2$ may occur to bring about in the equilibrium state at about 1900 K. If the recombination were completed rapidly enough, the nitrogen content of weld metal would be the same as in the equilibrium state. However, the incomplete recombination of atomic nitrogen may allow the presence of excess atomic nitrogen gas (N) on the surface of the weld pool in an arc welding process, which probably accelerates the reaction $N \rightarrow \underline{N}$ and results in higher nitrogen content of the weld pool than in the equilibrium state. Some amount of the excess nitrogen will be evolved from the weld pool during solidification process. These facts may be able to explain the large difference between the nitrogen content of the iron weld metal and the equilibrium solubility of the nitrogen in liquid iron as shown in Fig. 11.

An anomalous nitrogen absorption into weld metal is observed at low arc atmospheric pressure in the nitrogen welding atmosphere. It may be caused by the presence of larger amount of dissociated nitrogen at lower atmospheric pressure and the higher cooling rate of the weld metal because of smaller penetration area of weld metal.

In both $N_2$ - Ar and the $N_2$ - He welding atmospheres, the nitrogen contents increase considerably with increase of the nitrogen partial pressure up to about 0.005 MPa and then hold a constant value of about 0.06 wt% above 0.005 MPa. The constant nitrogen may be caused by a balance between the nitrogen absorption into the molten weld pool below an arc column and the nitrogen evolution from the molten pool on the other part.

The nitrogen content in the $N_2$ - $H_2$ welding atmosphere is smaller than in the other welding atmospheres. Molecular hydrogen ($H_2$) dissociates into atomic hydrogen (H) easily in an arc column unlike monatomic molecule such as Ar and He. This fact leads to better thermal efficiency and larger penetration, which results in slower cooling rate and larger amount of the nitrogen evolution. On the other hand, it is presumed that the evolution of the absorbed hydrogen promotes the nitrogen evolution from the molten pool.

## 5.2 Nitrogen absorption of mild steel weld metal during gas metal arc welding

Using bare mild steel electrode wire of 1.6 mm $\phi$, mild steel were welded in the welding atmospheres of $N_2$, $N_2$ - Ar, $N_2$ - $O_2$, $N_2$ - $CO_2$ and $N_2$ - $H_2$. The relationship between the nitrogen contents of mild steel weld metals and the nitrogen partial pressures in the welding atmospheres of various gas components is summarized in Fig. 12. It can be clearly understood that the nitrogen absorption into the weld metals depends appreciably upon the gases coexistent with nitrogen, even if welded at an identical partial pressure of nitrogen.

In the welding atmospheres of $N_2$, $N_2$ - Ar and $N_2$ - $H_2$, the nitrogen contents of the weld metals during gas metal arc welding show a similar tendency with those during gas tungsten arc welding. In the welding atmospheres of $N_2$ and $N_2$ - Ar, the nitrogen contents during gas metal arc welding seem to be lower than during gas tungsten arc welding. However, it is difficult to say its reason at present, because the chemical compositions of materials used and the welding conditions are different each other.

It is very interesting that the nitrogen content in the $N_2$ - $O_2$ welding atmosphere is remarkably higher than in $N_2$, $N_2$ - Ar and $N_2$ - $H_2$ welding atmospheres. However, the underlying mechanism for the effects of coexisting oxygen cannot be explained satisfactorily. The formation of NO or $NO_2$ in the atmosphere at high temperature, the retarding of the nitrogen evolution by the oxygen in the weld metal and others are supposed to be involved in the true mechanism.

The coexistence of carbon dioxide ($CO_2$) with nitrogen in the welding atmosphere causes an increase in the nitrogen content of the weld metal, but the effect of carbon dioxide is less than that of oxygen. In the $N_2$ - $CO_2$ welding atmosphere, the dissociated oxygen seems to play a role to increase the nitrogen content of the weld metal as in the $N_2$ - $O_2$ welding atmosphere.

6. Conclusions

In this paper, the absorptions of oxygen and nitrogen into the iron weld metal during arc welding are described using author's data mainly. In spite of the important problem, the fundamental study of the gas absorption into the weld metal is relatively poor in the literature. It may be caused by a difficulty of the experiment and the analysis on data, but there are many points which must be solved. Author expects that the following problems will be studied in the future.

(a) Thermodynamic data of metals and gases at higher temperature
(b) Relationship between welding state and equilibrium state
(c) Physical chemistry during welding
(d) Effect of the gas contamination on the weldability
(e) Utilization of the gas on the welding process
(f) Others

If some points are solved, it will contribute to welding engineering.

References

Wagman, D.D., Evans, W.H., Parker, V.B., Halow, I., Bailey, S.M. and Schuman, R.H. (1968) Selected Values of Chemical Thermodynamic Properties, Tables for the First Thirty-Four Elements in the Standard Order of Arrangement, NBS Technical Note 270-3, U.S. Dept. of Commerce, National Bureau of Standards, U.S.Government Printing Office, Washington.

Kelly, K.K. (1960) Contributions to the Data on Theoretical Metallurgy, VIII High-Temperature Heat-Content, Heat-Capacity, and Entropy Data for the Elements and Inorganic Compounds, Bulletin 584 Bureau of Mines, U.S.Government Printing Office, Washington.

Chipman, J. (1951) Basic Open Hearth Steelmaking, A.I.M.E., 621-690.

Pehlke, R.D. and Elliott, J.F. (1960) The Solubility of Nitrogen in Liquid Iron Alloys, 1.Thermodynamics, Trans. Met. Soc. AIME, 218-12, 1088-1101.

Kuwana, T. and Sato, Y. (1986) Absorption of Oxygen into Iron Weld Metal during Arc Welding, Trans. Japan Weld. Soc., 17-2, 124-132.

Kuwana, T. and Sato, Y. (1988) Oxygen Absorption and Oxide Inclusion of Iron Weld Metal during Arc Welding, Trans. Japan Weld. Soc., 19-2, 134-142. IIW, Doc. IX-1392-86(1986).

Kuwana, T. and Kokawa, H. (1988) The Nitrogen Absorption of Arc-Melted Iron and Stainless Steel, Trans. Japan Weld. Soc., 19-2, 92-99.

Kuwana, T. and Kokawa, H. (1986) The Nitrogen Absorption of Iron Weld Metal during Gas Tungsten Arc Welding, Trans. Japan Weld. Soc., 17-1, 20-26.

Kobayashi, T., Kuwana, T. and Kikuchi, Y. (1965) Arc Atmosphere and Nitrogen Content of Weld Metal, Welding in the World, 5-2, 58-73. IIW, Doc. XII-265-65(1965).

# TRANSIENT LIQUID INSERT METAL DIFFUSION BONDING OF NICKEL-BASE SUPERALLOYS

Y. NAKAO*, K. NISHIMOTO*, K. SHINOZAKI* and C.Y. KANG**
* Department of Welding and Production Engineering, Osaka University
** Department of Metallurgical Engineering, Pusan National University

Abstract
Transient Liquid Insert Metal Diffusion Bonding (TLIM bonding) consists of three processes, viz., a dissolution process of base metal, an isothermal solidification process and a homogenizing process. These processes are considered theoretically based on results of current researches.

Advanced TLIM bonding process using an amorphous and an alloy powder sheet is also presented. In this new process, both the time necessary to complete the isothermal solidification process and that for homogenizing process are shortened compared with those of conventional process. The mechanisum to shorten TLIM bonding process is discussed in detail.

Application of TLIM bonding process to advanced materials such as ODS alloy and a single crystal alloy is also presented.
Key words: TLIM bonding, Dissolution process, Isothermal solidification process, Homogenizing process, Diffusion control, Advanced TLIM bonding, Advanced materials, TLP bonding.

## 1. Introduction

Ni-base cast superalloys are a class of materials that are generally used hottest parts in aircraft or land-base gas turbines. As fair amount of $\gamma'$ forming elements such as Al, Ti, Nb, Ta is contained, these alloys are very sensitive to hot cracking during fusion welding(Thamburaj R., Wallace W. and Goldak J.A., 1983).

Transient Liquid Insert Metal Diffusion Bonding (TLIM bonding) has been recently developed to join hot cracking susceptible Ni-base cast superalloys, DS alloys, single crystal alloys and ODS alloys, and called 'TLP bonding' or 'Activated diffusion bonding'(Duvall D.S. Owczarski W.A. and Paulonis D.F.,1974).

In this process a liquid film temporarily forms at bonding region and solidifies isothermally due to interdiffusion of elements for depressing melting point of filler metal. As the result, the solidified bonding region consists of a primary solid solution similar in composition to base material. Consequently, the excellent joint quality can be expected without any fusion welding defects and any brittle phases. Therefore, TLIM bonding has been applied for bonding of Ni-base superalloys.

The bonding process of TLIM bonding is composed of the following three processes. Namely, the first one is the dissolution process of base metal adjacent to the liquefied insert metal. The second one is the isothermal solidification process and the last one is the homogenizing process of alloying elements. It is essentially important to make clear these processes theoretically in order to understand the mechanism of TLIM bonding.

In this report, theoretical considerations on the bonding mechanisms of this process are presented and application of this process to advanced Ni-base superalloys such as ODS alloy and single crystal alloy is also described.

Newly developed TLIM bonding process for wide-clearance bonding proposed by Nakao Y. etal.(1990a) is presented.

## 2. Bonding mechanism

### 2.1 Dissolution process of base metal

Nakao Y. etal.(1989) have modelled this process based on Nernst-Brunner theory(Moelwyn E.A.-Hughes,1947). On the other hand, North T. H.(1989a) has carried out finite difference modelling of this process. First, analytical results of the process by Nakao Y. etal. is outlined.

Kinetic studies have been carried out to make clear the dissolution phenomenon of base metal into liquid insert metal during TLIM bonding of MM007(Ni-10.2Co-7.98Cr-6.17Mo-6.01Al-4.16Ta-1.34Hf-0.09C-0.016B) using MBF-80(Ni-15.5Cr-3.7B) amorphous insert metal. The isothermal phase transfer from solid to liquid can be broken up into two consecutive steps. The first is the interface reaction in which atoms go from the solid into the liquid phase. The second is the diffusion process in which the solute atoms diffuse from the interface into the bulk liquid. The dissolution phenomenon of base metal into liquid insert metal was analysed based on the Nernst-Brunner theory which was formulated as follows.

$$n=n_s[1-\exp(-(KA/V)t)] \quad (1)$$

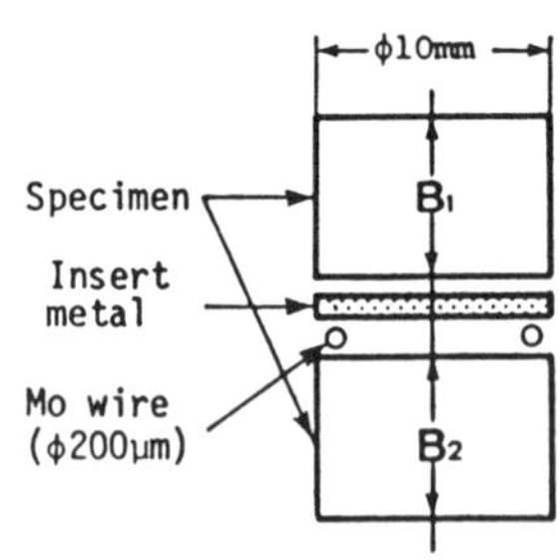


Fig.1 Specimen for bonding.

where $n_s$, n; solute concentrations at saturated state and at spontaneous time, t respectively, K; dissolution rate constant, A; surface area of the solid and V; volume of the liquid. When n and $n_s$ are expressed in weight %, $C_t$ and $C_s$ respectively, eq.(1) can be changed to eq.(2). $C_t$ and $C_s$ of B can be given by eq.(3) in the case of the joint illustrated in Fig.1.

$$dc/dt=K(A/V)(C_s-C_t) \quad (2)$$

$$C_t=(ph/(x+ph))C_i$$
$$C_s=(ph/(x_s+ph))C_i \quad (3)$$

where $C_i$; initial concentration of B in liquid insert metal, p; $(\rho_l/\rho_s)$, $\rho_l$, $\rho_s$; density of liquid insert metal and base metal, respectively, 2h; initial width of liquid insert metal and x, $x_s$; dissolution width of base metal at spontaneous time, t and at saturated state, respectively. Eq.(4) is given by eq.(2) and (3).

$$Kt=h[\ln(x_s(x+ph)/ph(x_s-x))]=P \quad (4)$$

where P; dissolution parameter. The linear relationship between holding time, t and dissolution parameter, P can be expected during dissolution process of base metal from eq.(4). Figure 2(a) indicates the relationship between holding time, t and P. As predicted from eq.(4), there is a good linear correlation between holding time, t and P. It is concluded that the dissolution phenomenon of base metal

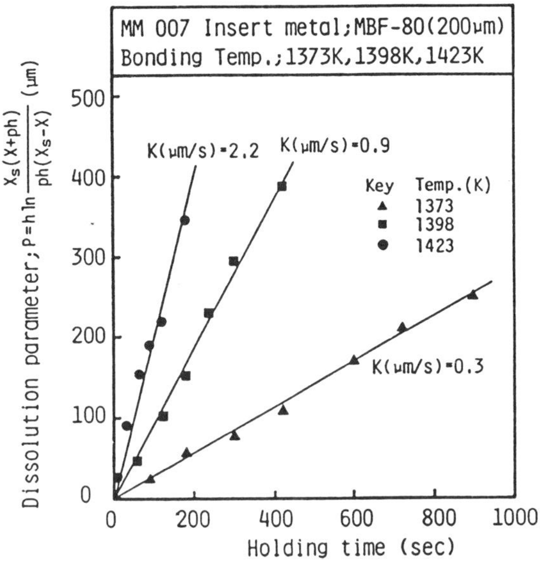

(a)

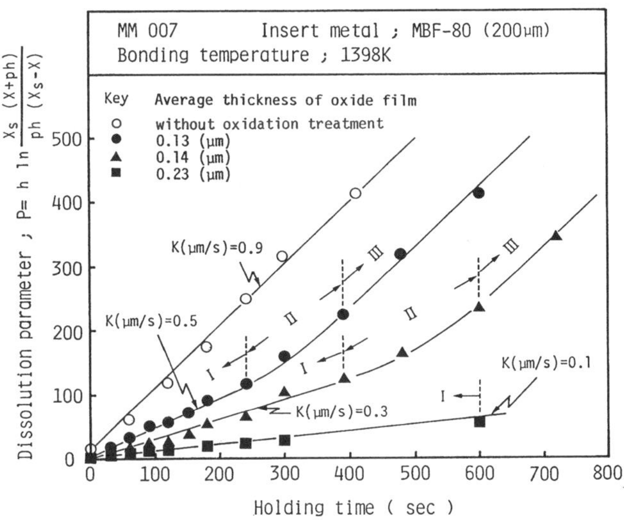

(b)

Fig.2 Relation between dissolution parameter P and time, t.
(a) no oxidation, (b) oxidation

into liquid insert metal can be explained by Nernst-Brunner theory.

In case of mating surfaces of base metals being oxidized, oxide films disturb dissolution of base metal into liquid insert metal. Figure 2(b) indicates effects of thickness of oxide film and holding time on dissolution parameter,P. The dissolution process consisted of three stages as shown in Fig. 2(b). In stage I, dissolution rate constant,K values are apparently constant and smaller than that value without oxidation treatment. When a diffusion couple with an insert metal is heated up to bonding temperature, the insert metal melts and after a while, liquid films are formed at the interfaces between base metals and oxide films on account of B atoms diffusion through oxide films to the interfaces. However, oxide films exist almost parallel to solid-liquid interfaces and those morphologies are nearly planner. K value decreases gradually in proportion to the thickness of oxide film ,because oxide film prevents B atoms diffusion from liquefied insert metal to base metal-oxide film interface. In stage III, oxide films decompose to fragments and as the result, K value of oxidized base metal becomes equal to that of non-oxidized base metal. Stage II is the transient stage on K value.

It is concluded from Fig. 2(b) that oxide film hinders the disolution reaction between base metal and liquefied insert metal. Oxides are finally swept up to grain boundaries in the bonded region during isothermal solidification.

North T.H. has carried out numerical analysis on the dissolution process of base metal during TLIM bonding of Ni using Ni-11P and Ni-14Cr-10P insert metals, assuming that this process is controlled by solute diffusion in liquid and solid. In this model the effect of heating rate between the insert metal melting temperature(Tm) and the brazing temperature(Tb) on the dissolution phenomenon of the base material for both constant and unrestrained liquid width situations. More advanced modelling of dissolution during TLIM bonding is achieved in this model.

2.2 Isothermal solidification process

Several researchers studied isothermal solidification process during TLIM bonding. First, analitical results of the process by Nakao Y. etal.(1989) is outlined. As shown in a schematical Ni-X binary phase diagram(Fig.3(a)), when the bonding pair with an insert metal of which X content is Ce is heated to the bonding temperature, $T_B$, a liquid film is formed at the bonding region. Shaded area of Fig.3(b) shows such a liquid film and a schematic X content profile at the bonding region is indicated in Fig.3(c). At the bonding temperature, $T_B$, a liquid phase of which X content is $C_1$ is in equilibrium with a

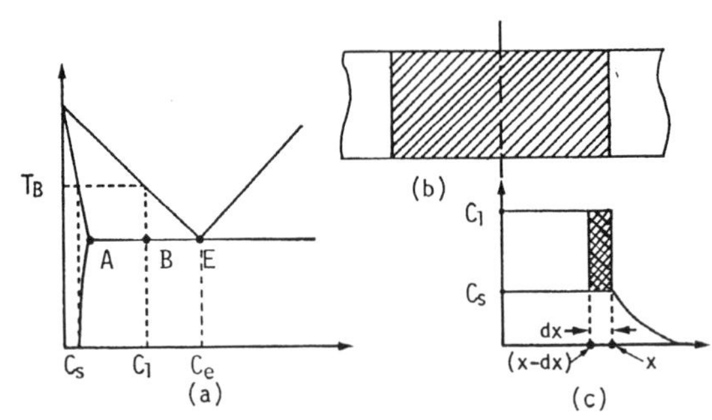


Fig.3 Schematic diagrams of Ni-X binary system and concentration profile.

solid phase of which X content is $C_s$. Consequently, at the interface between the liquid phase and the solid phase, both X contents of the liquid phase and the solid phase should be kept constant. If X atoms diffuse into the solid phase from the liquid phase, the solid-liquid interface must move to the minus direction of x axis to make X content in the liquid phase constant, $C_l$.

At the solid-liquid interface, the concentration gradient of X in the solid phase is assumed to be constant even if the solid-liquid interface travels. And as the liquid film was very thin comparing with the length of base metal, one dimensional diffusion model was adopted. In this model, the thickness of the liquid film, 2x at arbitrary time, t is given by eq.(5).

$$2x=2h-(4C_s/\sqrt{\pi}V_s)(C_l/V_l-C_s/V_s)^{-1}\sqrt{Dt} \qquad (5)$$

where $V_l$ is molar volume of the liquid phase at $T_B$, $V_s$ is molar volume of the solid phase at $T_B$, 2h is initial thickness of the liquid phase at arbitrary time, t, D is diffusion coefficient.

The complete time($t_f$) of the isothermal process can be estimated by eq.(6).

$$\sqrt{t_f}=(4C_s/\sqrt{\pi}V_s)^{-1}(C_l/V_l-C_s/V_s)x \; 2h/\sqrt{D} \qquad (6)$$

As it was a little difficult to determine the solid-liquid interface at room temperature, the thickness of Ni-X eutectic was measured in this experiment and the eutectic thickness, $W_e$ was converted into the thickness of the liquid film, $W_l$ using eq.(7).

$$W_l=W_e(1+(BE/AB)(\rho e/\rho s)) \qquad (7)$$

where $\rho$ e; density of Ni-P eutectic and $\rho$ s; density of Ni solid solution. As predicted from eqs.(5) and (7), there are good linear correlations between reduced width of eutectic and square root of holding time at each holding temperature, in Ni-B and Ni-P systems, as shown in Fig.4. From the results shown in Fig.4 and eq.(5), each activation energy of B and P diffusion in Ni can be

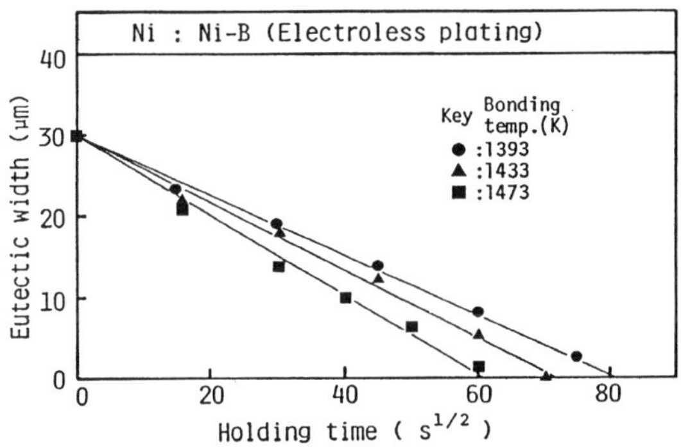


(a)

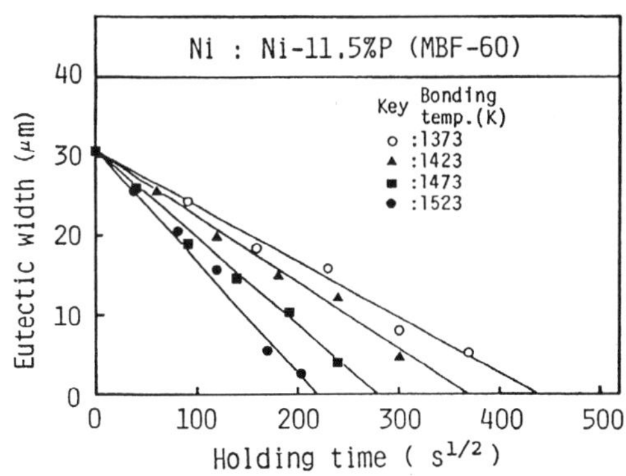


(b)

Fig.4 Effect of holding time at each temperature on eutectic width. (a) Ni-B, (b) Ni-P

estimated and each value is 226 kJ/mol and 284 kJ/mol, respectively. As the term, $(4C_s/\sqrt{\pi}V_s)^{-1}(C_l/V_l-C_s/V_s)$ in eq.(6) is nearly independent of the temperature in the temperature range from 1373K to 1523K on both Ni-B and Ni-P systems, the square root of $t_f$ is directly proportional to $2h/\sqrt{D}$ according to eq.(6).

As shown in Fig.5, there are good proportional relations between $t_f^{1/2}$ and $2h/\sqrt{D}$ in both Ni-B and Ni-P systems. The disappearance time of liquid phase at bonding region can be estimated by eq.(6) in both Ni-B and Ni-P systems. Based on the experimental results mentioned already, it was made clear that the isothermal solidification process of liquid film during TLIM bonding was controlled by the diffusional process of dipressant element in the base metal.

Onzawa T. etal.(1978) also studied on the isothermal solidification process and they estimated the time required for completion of this process($t_f$) by following equation.

$$t_f=0.19(Xm^2/Ds)(Cl/Cs)^2 \quad (8)$$

where Xm; initial half width of the liquid at isothermal solidification, Ds; the diffusibity of the melting point depressant element in the solid, Cl; the consentration of the depressant element in the liquid, Cs; the equilibrium concentration of the depressant element in the solid at the solid-liquid interface.

They concluded that eq.(8) was valid to evaluate $t_f$ for the isothermal solidification process during TLIM bonding of Alloy 713C using a Ni-15Cr-3.5 B insert metal.

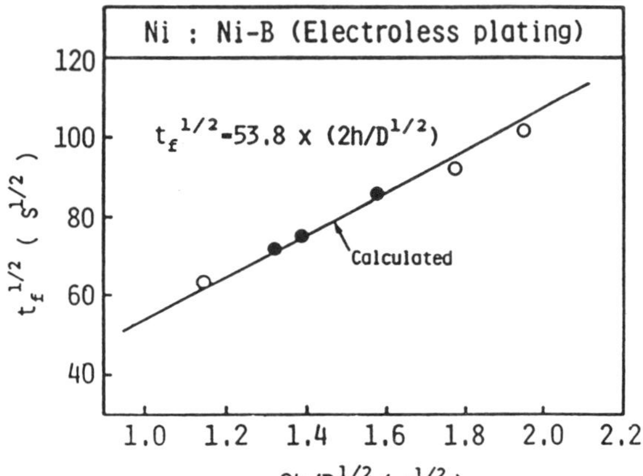


(a)

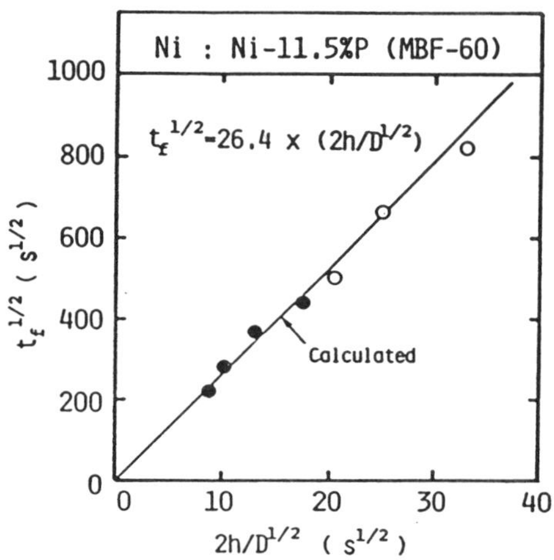


(b)

Fig.5 Comparison between calculated values of $t_f^{1/2}$ and experimental ones.
(a) Ni-B, (b) Ni-P

North T.H.(1989b) carried out a modelling of isothermal solidification process during TLIM bonding by a finite difference method using Ni-14Cr and Ni-20 Cr base metals with Ni-11P and Ni-14Cr-10P insert metals. The modelling was carried out for both variable and constant liquid-width situations. The effect of heating rate between the insert melting temperature (Tm) and the brazing temperature(Tb) on isothermal solidification process was also researched. North T.H. succeeded to achieve a more advanced modelling for the process.

2.3 Homogenizing process

The mechanical properties and chemical and physical properties of a TLIM-bonded joint can be equalized to those in the base metal after homogenizing process. Therefore, homogenizing process has very important roles in TLIM bonding of Ni-base superalloys and hence the relevant equation which can estimate the suitable homogenizing condition has been required in fabrication.

Modellings of the homogenizing process were carried out by Nakao Y. etal.(1990b) and Onzawa T. etal.(1978).

First, Modelling by Nakao Y. etal. is outlined. The time dependency on the concentration of alloying elements in the bondinhg region during homogenizing process was estimated using the least square method. An experimental eq.(9) was derived based on an exponential function modifying error function.

$$Ct=A_1+(100-A_1)(A_2t^{1/2})/(A_3t^{1/2}+A_4)\exp(-A_5/t) \qquad (9)$$

Experiments were carried out using MM007 base metal and MBF-80 insert metal. Table 1 presents the coefficients from A1 to A5 in eq. (9) and the correlation coefficients between the experimental values and calculated ones at each bonding temperature.

Table 1 Summarized coefficients and correlation coefficients of Al,Ti and Cr for eq.(9).

| Element | Temperature (K) | Coefficient $A_1$ | $A_2$ | $A_3$ | $A_4$ | $A_5$ | Correlation coefficient |
|---|---|---|---|---|---|---|---|
| Al | 1373 | 24.2 | 0.271 | 0.249 | 66.0 | 9000 | 0.998 |
| | 1423 | 39.1 | 0.635 | 0.420 | 82.0 | 2000 | 0.996 |
| | 1453 | 45.7 | 0.748 | 0.783 | 92.0 | 1600 | 0.981 |
| Ti | 1373 | 34.3 | 0.524 | 0.583 | 66.0 | 12000 | 0.996 |
| | 1423 | 39.0 | 0.921 | 1.000 | 82.0 | 12000 | 0.986 |
| | 1453 | 45.7 | 0.920 | 0.971 | 92.0 | 500 | 0.967 |
| Cr | 1373 | 171.4 | 0.136 | 0.111 | 66.0 | 25000 | 0.989 |
| | 1423 | 157.5 | 0.179 | 0.129 | 82.0 | 6000 | 0.988 |
| | 1453 | 151.2 | 1.391 | 0.775 | 92.1 | 2800 | 0.988 |

The calculated concentration values correspond well with the experimental ones as shown in Table 1 and Fig.6(compare the solid curves with the experimental results). Eq.(9) can be applied for estimating the time for completion of homogenizing process under same homogenizing conditions as in Fig.6. However, its extension under other homogenizing conditions is hardly applicable. Therefore, the equation which can be widely applied under any conditions is required in fabrication.

It is difficult to consider theoretically the diffusion process in commercial alloys since these are multi-component systems. However,it is reasonable to regard that Ni-base commercial superalloys are Ni-X binary alloys from fabrication view point.

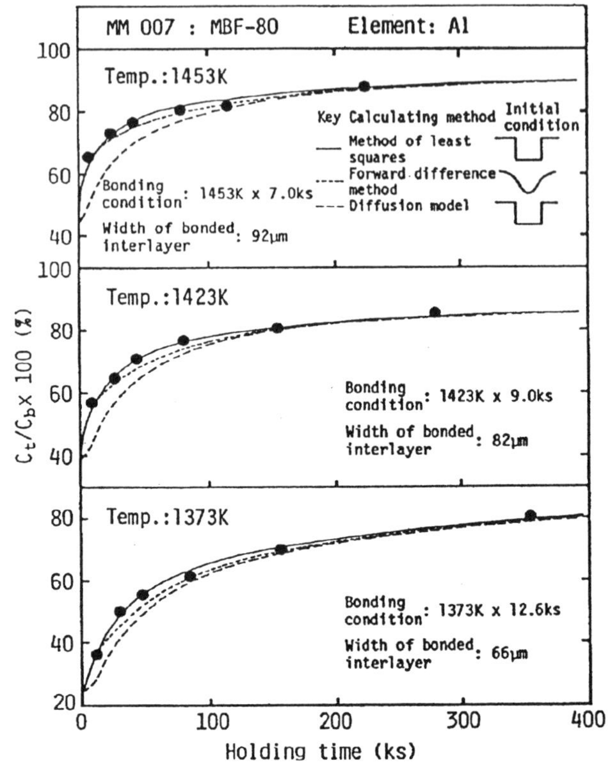


Fig.6 Calculated Al concentration and measured one at the center of the bonded interlayer of MM007 joint.

Figure 7 shows the schematic diagram of Al concentration profile at bonding region at the initial stage of homogenizing process ignoring Al diffusion into the bonded interlayer during the isothermal solidification. Assuming that the Fick's law with one-dimensional model is valid, the Al concentration, Ct at time, t is estimated analytically by applying eq.(10).

$$Ct=Co[1-\mathrm{erf}(h/2(Dt)^{1/2})]+Ci \qquad (10)$$

where 2h; width of bonded interlayer, Ci; initial concentration of Al in bonded interlayer, Cb; concentration of Al in base metal, Co; term (Cb-Ci) and D; diffusivity of Al.

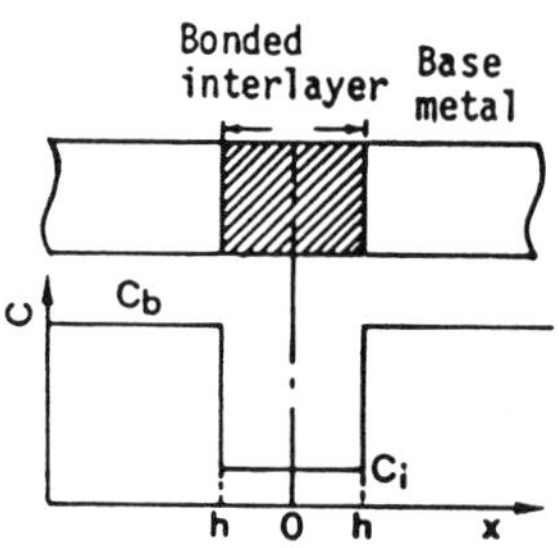


Fig.7 Concentration profile of element in the bonded region.

The diffusivity of Al in nickel employed is $1.87\times10^{-4}\exp(-268000/RT)$ ($m^2/s$). The base metal dissolution is considered during calculating. The broken curves in Fig.6 show the calculated values using eq.(10). They correspond with the experimental values except in the early stage of homogenizing process. The discrepancy between the experimental values and calculated ones may be due to assuming the uniform element concentration in the bonded interlayer at the initial stage of homogenizing process. For this reason, the calculation considering with the concentration distribution at the initial stage of homogenizing process was carried out using the finite difference analysis. The dotted curves in Fig.6 show the calculated values using the finite difference analysis. They correspond better with the experimental values even in the early stage of homogenizing process compare with those using eq.(10). Comparing the calculated concentration in Al by the least square method to those by analytical method in all base materials used (when Ct/Cb becomes 90%), it should be emphasized that the difference between both values was within 1%. It follows that eq. (10) can be used to estimate the holding time for completion of homogenizing process in fabrication.

Onzawa T. etal.(1978) also carried out a modelling of homogenizing process under same initial conditions as indicated in Fig.7, using Fick's second law. They estimated Al, Mo, Ti concentrations in the bonded layer of Alloy 713C joint with a Ni-15Cr-3.5B amorphous insert metal using eq.(10). They confirmed that the homogenizing process could be completed by annealing treatment at 1443K for 86.4 ks using eq.(10).

## 3. Advanced transient liquid insert metal diffusion bonding

### 3.1 Concept for advanced TLIM bonding

It has been reported that a clearance for bonding should be about 100 $\mu$m in fabrication of Ni-base superalloys by TLIM bonding method (Sakamoto A. etal.,1987). The calculation by eq.(10) resulted that the times required for completion of isothermal solidification, tif and sequent homogenizing process,thf were 48 and 2870 ks,respectively in case of bonding MM007 alloy with a bonding clearance of 100$\mu$m using a Ni-15%Cr-4%B insert metal at the bonding temperature of 1423 K (the time, thf was determined by the time for Al concentration at the center of the bonded interlayer corresponding with 90% of that at the base metal). Such long period during TLIM bonding process seems to be inefficient in fabrication as well as may decrease mechanical properties of base metal. For this reason, fabricators keen to be developed the new bonding techniques which shorten the period during TLIM bonding process keeping a wide bonding clearance of about 100$\mu$m.

Recently, a new TLIM bonding technique was developed, viz., alloying powder sheet whose chemical compositions were similar to those of a base metal was inserted with an insert metal during TLIM bonding as shown in Fig.8. In this technique, alloying powders do not melt down at the bonding temperature except at the interface to liquid insert metal, thereby two beneficial effects may be produced, viz., 1) decreasing the amount

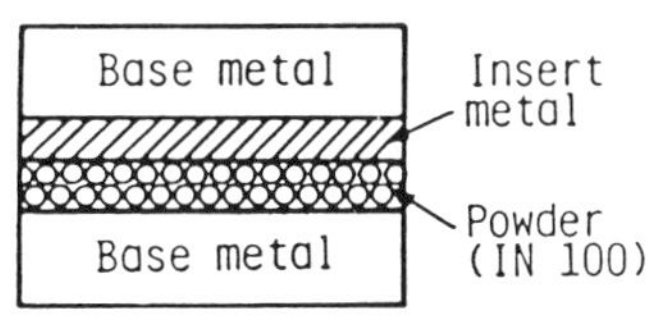

Fig.8 Schematic diagram of bonding pair with powder sheet.

of liquid phase in the bonded interlayer because of expelling excess melted insert metal by alloying powders and 2) accelerating element diffusion from alloying powders to bonding region through insert metal by the three-dimensional diffusion process.

These effects should shorten the times for completion of isothermal solidification and sequent homogenizing process.

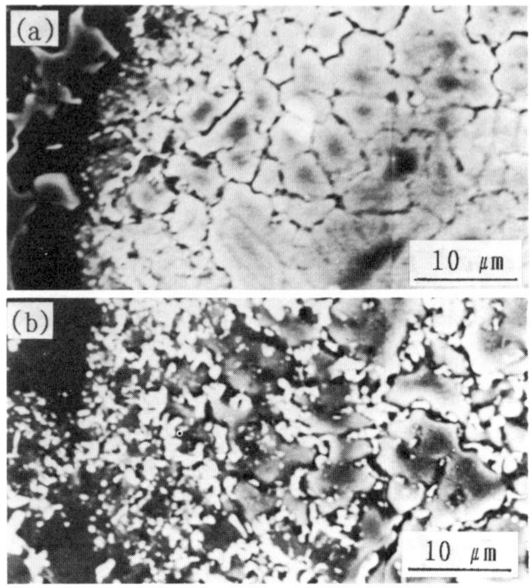


Fig.9 SEM microstructures of powders in the bonded interlayer.
(a) 0 s, (b) 40 s

## 3.2 Isothermal solidification process

Figure 9 shows the microstructures of alloying powders in the bonded interlayers of MM007 alloy joints at the holding time of 0 s and 40 s at 1398 K, respectively. The alloying powder is IN-100(average dia. 53$\mu$m) and the insert metal is MBF-80. At the time of 0 s, many fine particles are observed at the region adjacent to the insert metal/ alloying powder interfaces and not at the dendrite boundaries in alloying powders.
On the other hand, at the time of 40 s, many larger particles are clearly formed along the dendrite boundaries and also the widths of dendrite boundaries become wide. Such dendrite boundary morphology indicates that the local melting occurs at the dendrite boundaries. According to the analysis using EPMA and TEM technique, the particles along dendrite boundaries were identified borides as a type of $M_3B_2$.

Figure 10 shows the relation between the holding time and the region where borides appear in alloying powder as a function of the depth from the insert metal/alloying powder interface. Although the region of fine particles at the insert metal/ alloying powder interface is nearly independent of the holding time, the large boride region in alloying powder along the dendrite boundaries is dependent on the holding time. It should be pointed out that the

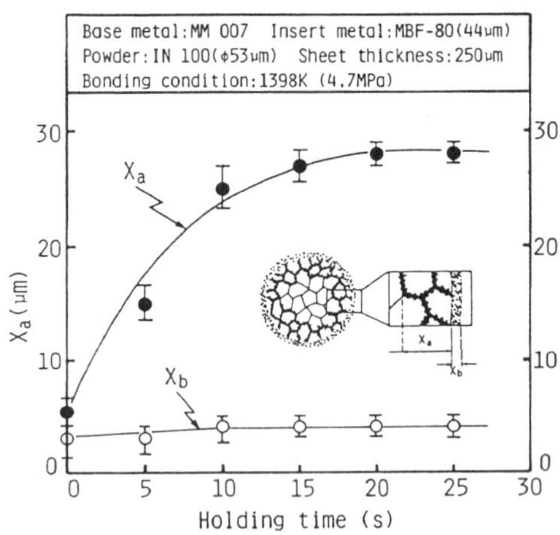


Fig.10 Effect of holding time on the boride forming region.

borides form at whole dendrite boundaries in alloying powders at the holding time of 20 s. This time seems to be too short for the boron atoms to diffuse into the alloying powders to form borides along the dendrite boundaries. Therefore, the formation of borides in alloying powders presumes to be controlled by liquid insert metal immersing into the dendrite boundaries in alloying powders.

On the other hand, the eutectic phases like those as indicated in Fig.9(a) form at the interpowder regions. They were considered to be liquid phases at the bonding temperature since they consisted of borides and Ni solid solutions. Figure 11 shows the relation between the holding time and the relative amount of the eutectic phase of a MM007 joint with alloying powders and that without alloying powders which is indicated by a solid linear line. The linear relation was estimated using eq.(5). The initial amount of liquid insert metal at bonding region without alloying powders was equalized to that in case of using alloying powders with a bonding clearance 100 $\mu$m in calculation. The amount of the eutectic phases decreases with the holding time and then becomes almost zero at the holding time of about 120 s at 1398 K, viz., isothermal solidification completes at this time. However, in case of no alloying powders, the time for completing isothermal solidification becomes to be about 48 ks even at 1423 K.

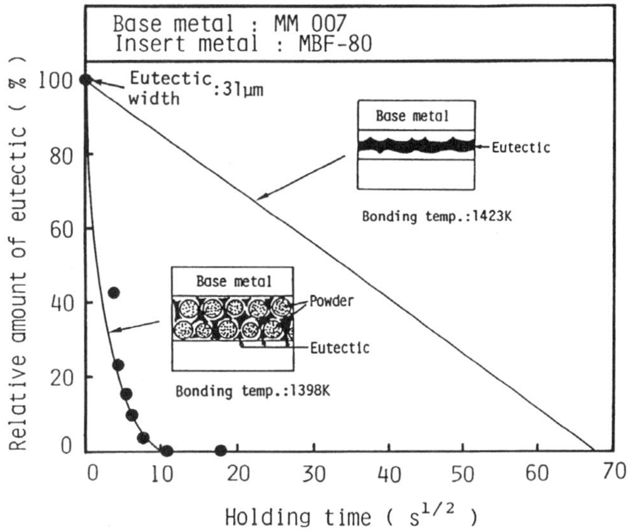


Fig.11 Effect of holding time on relative amount of eutectic.

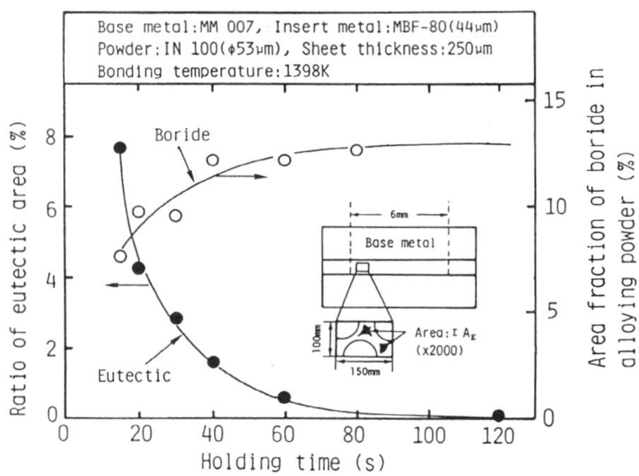


Fig.12 Effect of holding time on amounts of eutectic and boride.

Figure 12 also shows the relation between the holding time and the amount of borides at the dendrite boundaries in alloying powders and that between the amount of eutectic phases at the interpowder regions

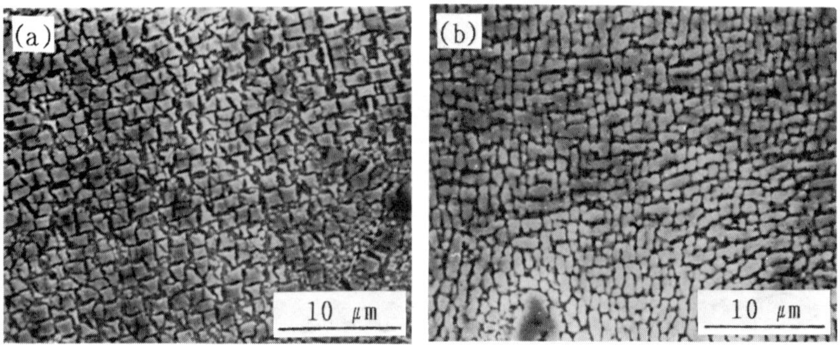


Fig.13 SEM microstructures of bonded interlayer of MM007 joint and base metal. (a) 1423 K x 86.4 ks, (b) base metal

and the holding time in a MM007 joint with alloying powders. Increasing the holding time increases the amount of borides while that decreases the amount of eutectic phases. The result follows that the isothermal solidification is advanced by liquid insert metal immersing into the dendrite boundaries in alloying powders and sequent crystalizing borides along the dendrite boundaries in alloying powders. The reason why the time, tif was too short using alloying powders may be related to this process. It should be emphasized that this process is quite different from the conventional process where the moving rate of the solid-liquid interface is controlled by diffusion of the melting point depressant elements into base metal.

3.3 Homogenizing process

Figure 13 shows the microstructure of the bonded interlayer in a MM007 joint homogenized at 1423 K for 86.4 ks and that of the base metal. It is clearly apparent that $\gamma'$ phase precipitation depends on the holding time and the microstructure of $\gamma'$ at the bonded interlayer homogenized at 1423 K for 86.4 ks is similar to that of base metal.

The hardness of bonded interlayer increases with the homogenizing time, viz., in relation to the amount of $\gamma'$ phases and the hardness for 86.4 ks corresponds well with that of base metal.

The experimental results

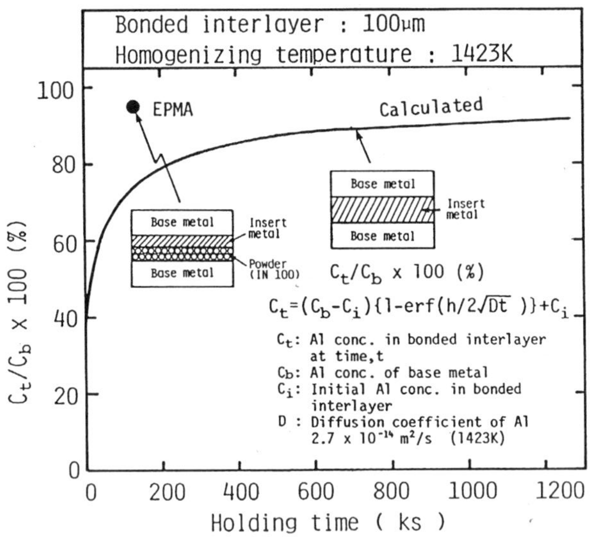


Fig.14 Effect of holding time on Ct/Cb. Ct; Al concentration at the center of bonded interlayer Cb; Al concentration of base metal

show that the holding time of 86.4 ks is enough to homogenize concentrations of TLIM-bonding joint using alloying powders (Fig.14). This homogenizing time, 86.4 ks is thirty times shorter in comparison to the calculated one of 2870 ks without using alloying powders as mentioned previously. The effect of alloying powder on shortening the homogenizing time may be due to elements diffusing from alloying powders to bonding region by three dimensional diffusion process.

From results of tensile tests of MM007 TLIM bonding joints using alloying powders and MBF-80 insert metals, it should be emphasized that MM007 joints fractured at base metal positions at the testing temperature of 1033 K and their tensile strengths at 1144 and 1255 K were about same as those of the base metal in spite of failure occurring at the bonded interfaces. These test specimens were bonded at 1423 K for 1.8 ks and subsequently homogenized at 1423 K for 86.4 ks. After that they were aged at 1173 K for 36 ks. The experimental results at 1255 K of Mar-M247 and Alloy 713C joints are same as those of MM007 joints.

From the experimental results, it should be useful to insert alloying powders to a wide bonding clearance joint during TLIM bonding in fabrication.

## 4. Application TLIM bonding process to advanced materials

Morimoto S. and Hirane T. (1986) tried to apply TLIM bondiong process to join an ODS alloy, MA 754(Ni-19.76Cr-0.99Fe-0.42Ti-0.36Al-0.07C-0.56$Y_2O_3$).

Four kind of ribbons(A,B, C,D) were used as insert metals of which chemical compositions were Ni-20Cr-4B(ribbon A),Ni-21.8Cr-4.5B (ribbon B),Ni-21.4Cr-3.3B-2.9Si(ribbon C) and Ni-20.4 Cr-3.2B-7.1Ge(ribbon D), respectively. These ribbons are amorphous ribbons and the average thickness is about 20$\mu$m. Bonding conditions are as follows; bonding pressure; 0.1 MPa, bonding temperature; 1393, 1433, 1473 K, bonding time; 1.8, 3.6, 18, 72 ks, vacuum; $6.7 \times 10^{-4}$ Pa. Specimens for tensile tests and creep

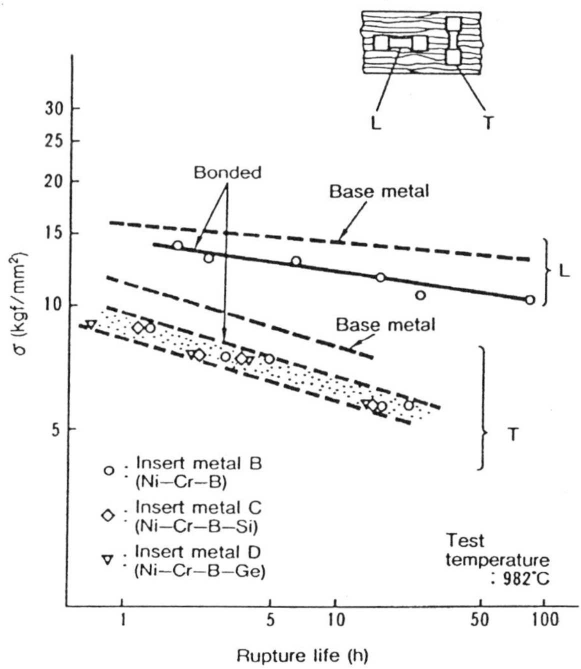


Fig.15 Creep rupture properties of ODS alloy, MA 754 bonded joints.

rupture tests were bonded at 1473 K for 36 ks. Their main results are as follows;

(1) $Y_2O_3$ particles and chemical compounds of Al and Ti agglomerated at grain boundaries in the bonded interlayer, however no bonding defects such as pores were found.

(2) Although joints bonded using ribbons A, B and C as insert metals fractured in base metals at the temperature higher than 773 K, joints bonded using ribbon D fractured at bonded interlayers at 773 K. However, such joints fractured in base metals when the test temperature was higher than 1073 K.

(3) Creep rupture lives of joints at 1255 K were about 85% of that of base metal (Fig.15).

Funamoto T. and Hirane T.(1990) tried to bond a single crystal alloy of which chemical composition is Ni-10.5W-9.0Cr-5.7Al-3.2Ta-1.2Ti-1.0Mo by TLIM bonding process. Bonding pairs were bonded at 1473 K for 3.6 ks in vacuum using Ni-4.1B-2.9Si amorphous ribbons (thickness; 16 $\mu$m) as insert metals. These joints fractured at needle like precipitates of which main element was W as bonded condition. The precipitates located at the interfaces between base metal and bonded interlayer.

The effect of homogenizing treatment on tensile properties of the joints is indicated in Fig.16. Homogenizing treatment for 36 ks at the temperature higher than 1473 K was effective to prevent precipitation of brittle W-phases and as the result, joints fractured at bonded interlayers or base metals, and mechanical properties were improved as illustrated in Fig.16.

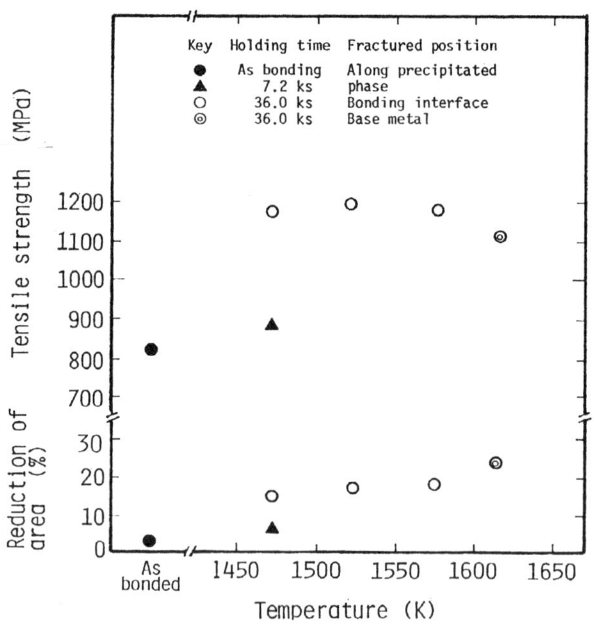


Fig.16 High temperature tensile properties of single crystal alloy (Ni-10.5W-9.0Cr-5.7Al-3.2Ta-1.2Ti-1.0Mo) bonded joints.

## 5. Conclusions

In Transient Liquid Insert Metal Diffusion Bonding (TLIM bonding) process the solidification occurs isothermally. As the result, the bonded region consists of the primary solid solution whose chemical composition is similar to that of base metal and the posibility of the formation of brittle phases can be avoided unlike brazing. TLIM bonding process also has the another advantage of not requiring rather higher pressure like solid state diffusion bonding. Weld defects shch as hot crackings are also avoided using this process. Concequently TLIM bonding process has currently used to join superalloys and some researchers have tried to apply this process for joining of advanced materials such as DS alloys, eutectic alloys, single crystal alloys, ODS alloys and MMC.

In this report, theoretical approaches to TLIM bonding process are described at first. Then a new TLIM bonding process for wide-clearance bonding is presented. Finally, The application of TLIM bonding process for joining an ODS alloy, MA 754 and a single crystal alloy is presented.

Main conclusions are as follows.

(1) On theoretical consideration for TLIM bonding process, three main processes such as dissolution process, isothermal solidification process and homogenizing process should be researched theoretically in multi-component system because superalloys consist of many kind of alloying elements. Since the heating rate from the melting temperature of insert metal to the bonding temperature is one of very important parameters to control three processes as mentioned above, the effect of the heating rate on the TLIM bonding process should be also studied theoretically.

(2) New bonding techniques using liquid film for wide-clearance bonding should be developed and researched theoretically because the clearance of bonding pair will be possible to become wider on account of thermal strain caused by inhomogenious heating in fabricating industrial parts such as gas turbine parts.

The authers proposed to use powder sheet with insert metal to shorten the time for homogenizing treatment and confirmed that this new technique was very useful to bond wide-clearance bonding pairs. However, theoretical consideration is not enough to understand whole bonding process. The morphology and size of powder, kind of powder and the thickness of powder sheet are also factors to control mechanical properties of bonded joints. The effects of these factors on properties of bonded joints should be studied.

(3) TLIM bonding of advanced materials should be researched theoretically focusing the influence of bonding process on strengthening mechanisms of advanced materials and also studied based on engineering view point.

References


Thamburaj R., Wallace W. and Goldak J.A. (1983) Post-Weld Heat-Treatment Cracking in Superalloys. Inter. Metals Reviews, 28-1, 1-22.

Duvall D.S., Owczarski W.A. and Paulonis D.F. (1974) TLP Bonding ; a New Method for Joining Heat Resistant Alloys. Weld. J., 53, 203-214.

Nakao Y., Nishimoto K., Shinizaki K., Kang C.Y. and Shigeta H.(1990a) Advanced Transient Liquid Insert Metal Diffusion Bonding Using Alloying Powder. to be published.

Nakao Y., Nishimoto K., Shinozaki K. and Kang C.Y. (1989) Theoretical Research on Transient Liquid Insert Metal Diffusion Bonding of Nickel Base Alloys. Trans. Japan Weld. Soc., 20, 60-65.

Moelwyn E.A.-Hughes (1947) The Kinetics of Reaction in Solution. Clarendon Press, Oxford, 374-377.

North T.H. (1989a) Private communication.

Onzawa T., Mitsunaga Y., Suzumura A., Tamura H. and Morikawa M. (1978) Structure and Mechanical Properties of TLP bonded Heat Resistant Superalloy Inco. 713C. J. Japan Weld Soc., 47, 161-166.

North T.H. (1989B) Private communication.

Nakao Y., Nishimoto K., Shinozaki K., Kang C.Y. and Shigeta H.(1990b) Analysis of Homogenizing Process of Alloying Elements on Transient Liquid Insert Metal Diffusion Bonding. to be published.

Sakamoto A., and Fujiwara C. (1987) Preprints National Meeting of Japan Weld. Soc. 41, 48.

Morimoto S. and Hirane T. (1986) Liquid Phase Diffusion Bonding of Oxide Dispersion Strengthened Alloy MA 754. Report of the 123rd Committee on Heat Resisting Metals and alloys, Japan Soc. for Promotion of Science, 27, 255-263.

Funamoto T. and Hirane T. (1990) Private communication.

# A NEW MODELING OF WELD SOLIDIFICATION AND ITS APPLICATION TO CRACK SUSCEPTIBILITY FOR STAINLESS STEELS

F. MATSUDA and Jong-Bong LEE*
Welding Research Institute, Osaka University, Japan
* Presently with RIST, Republic of Korea

Abstract

Sequence of solidification front was investigated for various stainless steel weld metals by means of quenching in liquid-tin. Using the solidification sequence, a new two dimensional solidification model of cellular dendritic growth was proposed in order to study the weld solidification phenomena of stainless steel. Then, calculations of fraction solid against temperature decrease during solidification and segregation of the solutes within the cellular dendrites were done by means of finite difference method. The results calculated were compared with the experimental results which were obtained by quenching for each stainless steel. As a result, it was concluded that the modeling in this study is valid enough in order to understand the weld solidification. Moreover, the modeling was applied to the estimation of the Solidification Brittleness Temperature Range (BTR) in fully austenitic and ferritic stainless steel welds. The magnitude of the BTR and the detrimental roles of each element in the BTR were calculated. Furthermore, the detrimental effect of nitrogen in the BTR of duplex stainless steel were estimated by calculation. The predictions calculated were experimentally confirmed by the Trans-Varestraint cracking test.

Key words : Solidification, Weld metal, Hot Cracking, Computation, Stainless steels.

## 1. Introduction

It is well known that solidification crack susceptibility of weld metal in stainless steel is strongly depended on solidification mode and microsegregation. Therefore, many researchers wished to observe solidification mode and then tried to reveal the solidification sequence by quenching in water during welding. However the detailed shape of dendrite tip near solidification front was not clear enough yet, especially for ferritic stainless steel, because of insufficient cooling rate during water quenching. Kou(1982) reported that the cellular dendrite of ferritic stainless steel can be observed by liquid-tin quenching during welding.

In order to have more detailed information of the solidification tip of weld metal, the authors tried to the liquid-tin quenching for various stainless steel base and weld metals during GTA welding. Then the detailed investigations of solidification sequence were done for stainless steel weld metals whose residual ferrite contents in room temperature was varied from 0 to 100%. This intensive quenching revealed the many interesting results in relation to solidification of stainless steels. These findings urged the authors to make a new modeling of cellular dendritic growth mode which was matched with weld solidification revealed by liquid-tin quenching. Many different modelings of dendritic solidification have been proposed so far. However, all the modeling proposed until now did not satisfy the findings in this study.

In this paper, therefore, another type of two dimensional modeling of cellular dendritic growth during weld solidification was proposed by using the behaviors obtained by liquid-tin quenching. The usefulness of the modeling was checked by EPMA analysis on microsegregations during and after weld solidification. Moreover, the calculation of the Solidification Brittleness Temperature Range (BTR) was done based on the observation that the lower temperature limit of BTR corresponds to the temperature at which residual liquid enriched with impurities solidifies completely. Furthermore, the ranking of harmful elements to the BTR was predicted for various stainless steel weld metals. Finally the reason of high crack susceptibility in duplex stainless steel was persured and confirmed experimentally.

## 2. Modeling and numerical analysis of cellular dendritic growth mode and microsegregation

### 2.1 Solidification sequence revealed by liquid-tin quenching

In order to investigate a clear microstructural change near solidification front during weld solidification, liquid-tin (230℃) quenching method was used during GTA welding without filler metal. Stainless steels used for quenching were SUS(AISI)310S, 316L, 304L, 329J2L, 430 and 447J1 of base metals with 1.5 to 3.0mm in thickness. GTA welding conditions were used 55-90A, 12-14V and 50mm/min depending on the sheet thickness, in order to get a full penetrated weld bead. Temperature distribution on the surface along weld center was measured just before quenching with a W-5%Re/W-26%Re thermocouples.

As a result liquid-tin quenching method was much higher in the cooling rate during quenching than water quenching method and revealed more detail of the dendritic substructure for not only primary austenitic but also primary ferritic structures.

Figure 1 shows an example of microstructures within BTR of SUS310S 2.0 mm thick fully austenitic stainless steel. Cellular dendrites of the austenite can be obviously seen from the solidification front to the lower temperature limit of BTR. With observing the microstructure closely, it is guessed that the solid bridge was formed around mark "A" near the solidification front due to the mutual contact of

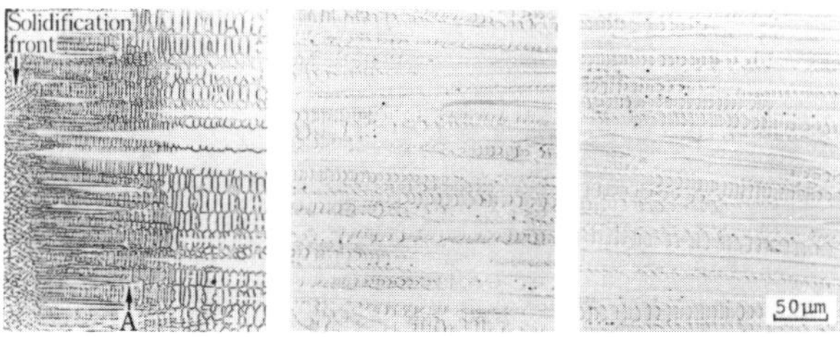


(a) 1670K (b) 1600K (c) 1550K

Fig.1 Quenched solidification microstructures within BTR (1670～1550K) of SUS310S

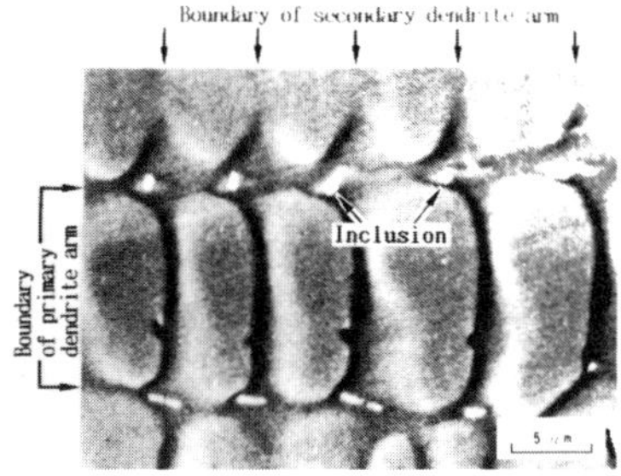


Fig.2 SEM microstructure near 1550K

secondary dendrite arm. A typical magnified microstructure near the lower temperature limit of BTR is given in Fig.2. Inclusions are observed at the location where the primary dendrite arm boundary intersects with the secondary dendrite arm boundary. This means that the residual liquid droplets are confined finally to the intersection due to development of the solid bridge. The quenched solidification microstructure of SUS329J2L which solidifies with only ferrite is given in Fig.3. Cellular dendrites are clealy observed near the solidification front. However, it was rapidly obscured with decreasing the temperature and was barely visible at the lower temperature limit of BTR. In addition, it is interesting to note that the secondary dendrite arm boundary was rapidly obscured compared with primary dendrite arm boundary.

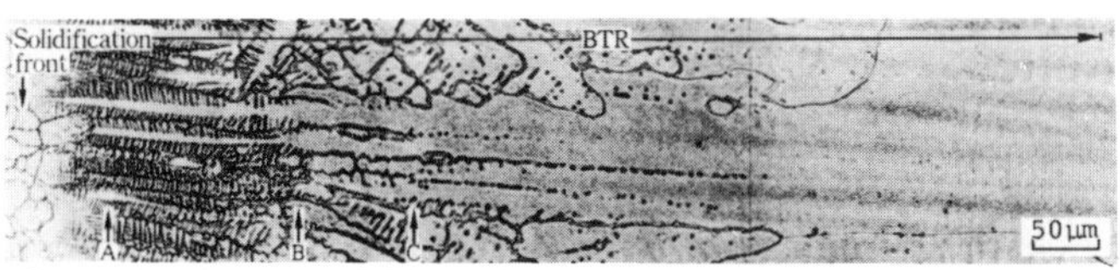


Fig.3 Quenched solidification microstructure of SUS329J2L

As a result, the authors have made the schematic representations of the weld solidification sequence for each steel as Fig.4, which are based on the quenched solidification microstructures of various stainless steels.

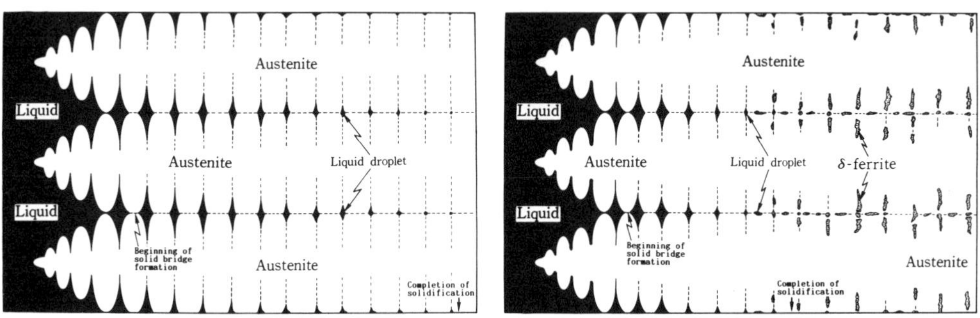


(a) SUS310S (b) SUS316L ( δ =2%)

Fig.4 (Continued)

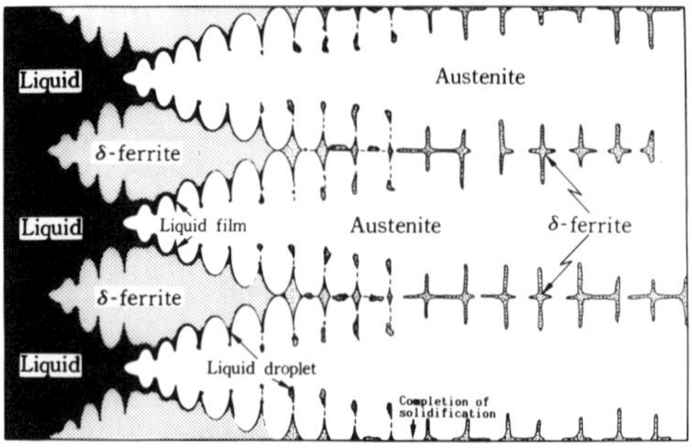


(c) SUS316L (δ =6%)

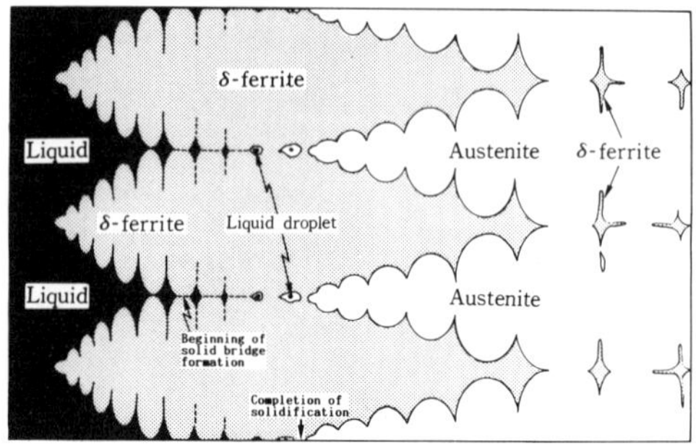


(d) SUS304L

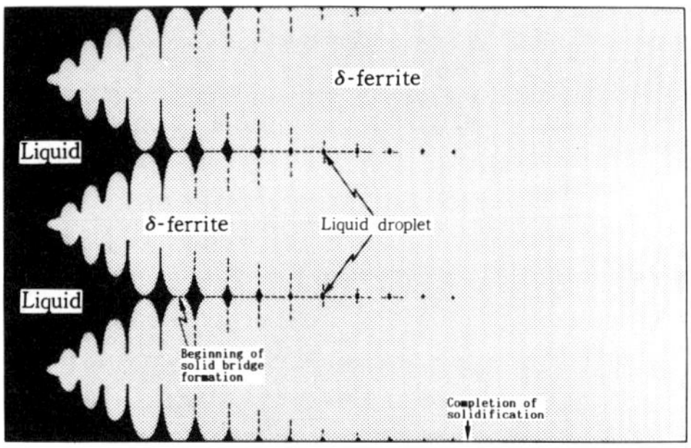


(e) SUS329J2L, SUS430

Fig.4 Schematic illustrations of solidification sequence based on quenched solidification microstructures (Reducing in the transverse direction)

### 2.2 Modeling and numerical analysis

#### 2.2.1 Two dimensional model of cellular dendrite

Figure 5 explains the two dimensional model used. The solidification sequence which solidifies with single phase is schematically illustrated in (a), although its right half is reduced considerably in the transverse direction for the purpose of simplification. Solid bridges between secondary dendrite arms growing from the neighbouring primary dendrite arms are formed at a very high temperature near solidification front. Due to developing of the solid bridge, residual liquid droplets are confined to the intersectional site of the primary and the secondary dendrite arm boundary. Rectangle OHPW in (a) is one volume element for numerical analysis. The lengths $\overline{OH}$ and $\overline{OW}$ correspond to one halves of primary and secondary dendrite arm spacing, respectively. In this model, as shown in (b)-①, solidification in OHPW starts from O, and the solid-liquid interface advances at a rightangle to $\overline{WH}$. Solid bridges are formed at H and W at the stage shown in (b)-②. This means that the solid bridges in this model are formed at fraction solid of 50%. Then, the residual liquid is gradually confined to the region near P, where solidification finishes, as shown in (b)-③. Its detail is shown in (c) with coordinates x and y. The solid-liquid interface has a terraced shape. In the model used, the number of division $n_x$ in x direction is the same as the number $n_y$ in y direction in this paper. In this paper, $n_x \times n_y$ is named the total number of division $N_T$.

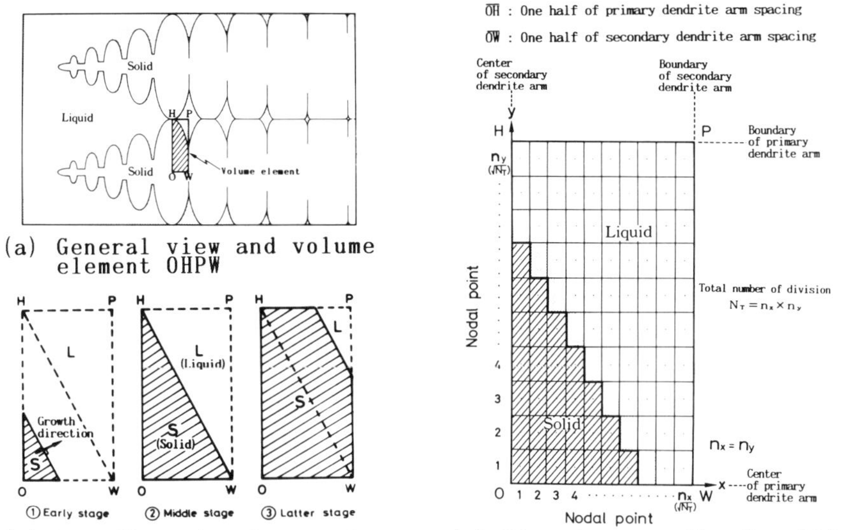


(a) General view and volume element OHPW

(b) Two dimensional growth in volume element

(c) Stepwise growth of solid-liquid interface

Fig.5 Growth characteristics of cellular dendrite based on quenched solidification microstructures and volume element proposed for calculation by two dimensional growth model

### 2.2.2 Assumptions for numerical analysis

Main assumptions adopted are as follows :

(1) Solute concentration in liquid is uniform. This might be roughly reasonable because of high diffusivity in liquid.

(2) Solute diffusion in solid has a temperature dependence, but has no concentration dependence.

(3) Local equilibrium always exists at the solid-liquid interface. In other words, solute is redistributed at the solid-liquid interface by equilibrium partition coefficient. Equilibrium partition coefficient is constant irrespective of temperature and concentration. Supercooling due to the curvature of solid-liquid interface and supercooling required for kinematics for advance of the solid-liquid interface are neglected.

(4) Melting point depression by solute, namely liquidus gradient is constant irrespective of temperature and concentration.

(5) There is no interaction between solute elements concerning diffusivity, partition coefficient, liquidus gradient.

(6) Coarsening of the secondary dendrite arm spacing near solidification front is neglected.

(7) Volume contraction accompanying solidification is neglected.

(8) Solid-liquid interface advances in a stepwise manner, which is mentioned in the next.

### 2.2.3 Procedures of numerical analysis

This problem is a moving boundary, and solved by the explicit finite difference method. Therefore, the most important problem is how to advance the solid-liquid interface. In the conventional modelings, the solid-liquid interface has been generally represented as a feature of parabolic growth, that is, it advances in proportion to square root of time. In this paper, however, the solid-liquid interface advances so that the liquidus temperature calculated may correspond to the actual temperature : Namely, the oblique lined region in Fig.5 (c) corresponds to solidified region, and suppose that solid-liquid interface advances suddenly by dotted zone. By this sudden advance the solute concentrations in liquid become higher than those in equilibrium with actual temperature ($T_A$). In other words, the liquidus temperature ($T_C$) calculated from the concentrations is lower than $T_A$. Therefore, the interface must be stopped temporarily until $T_C$ get the same value as $T_A$ by the solute diffusion in solid. Then, the interface advances again suddenly. Whole process is composed of the repeat of sudden advance and temporary pause.

Solute concentration in solid at the solid-liquid interface, $C_S(I,J)$, is always given by Eq.(1) irrespective of sudden advance or temporary pause.

$$C_S(I,J) = k \cdot C_\ell \quad \cdots\cdots \quad (1)$$

where k : equilibrium partition coefficient, $C_\ell$ : concentration in liquid, I and J are given by a fuction of the position of solid-liquid interface.

Solute concentration in solid at general nodal point (i,j), $C_S(i,j)$, during temporary pause is given by Eq.(2).

$$C_S(i,j)=C_{SB}(i,j) + D_S \cdot \Delta t/\Delta x^2\{C_{SB}(i+1,j)+C_{SB}(i-1,j) - 2C_{SB}(i,j)\} + D_S \cdot \Delta t/\Delta y^2\{C_{SB}(i,j+1)+C_{SB}(i,j-1) - 2C_{SB}(i,j)\} \quad \cdots\cdots \quad (2)$$

where $C_{SB}$ : solute concentration just before one time step $\Delta t$, $D_S$ : diffusion coefficient in solid phase, $\Delta x$, $\Delta y$ : dimension of a mesh, namely $\Delta x=\overline{OW}/n_x$, $\Delta y=\overline{OH}/n_y$, $\Delta t$ : time step $<(\Delta x \cdot \Delta y)^2/2D_S(\Delta x^2+\Delta y^2)$.

According to the assumptions (4) and (5), calculated liquidus temperature $T_C$ is given by Eq.(3) for all elements included.

$$T_C=T_L-\{m_C([\%C]_\ell-[\%C])+m_{Si}([\%Si]_\ell-[\%Si])+m_{Mn}([\%Mn]_\ell-[\%Mn])+ m_P([\%P]_\ell-[\%P])+\cdots\cdots\} \quad \cdots\cdots \quad (3)$$

where $m_X$ : liquidus gradient by solute X (K/wt.%), $[\%X]_\ell$ : concentration of solute X in liquid, $[\%X]$: initial concentration of solute X, $T_L$ : initial liquidus temperature of material.
Actual temperature $T_A$, is given by Eq.(4)

$$T_A=T_L-R_C \cdot t \quad \cdots\cdots \quad (4)$$

where $R_C$ is the cooling rate of weld metal.

### 2.2.4 Physical constants and other conditions used for calculation

Concentrations of C, Si, Mn, P, S, Cr, Ni and Mo were numerically analyzed simultaneously in this modeling, and their equilibrium partition coefficient k, liquidus gradient m, vibration factor $D_0$ and activation energy Q to calculate diffusivity in solid are listed in Table 1.

Table 1 Equilibrium partition coefficient k, liquidus gradient m, vibration factor $D_0$ and activation energy Q for diffusivity used in calculation

| Element | in Austenite | | | | in Ferrite | | | |
|---|---|---|---|---|---|---|---|---|
| | k | m (K/wt.%) | $D_0$ (mm²/sec) | Q (kJ/mol) | k | m (K/wt.%) | $D_0$ (mm²/sec) | Q (kJ/mol) |
| C | 0.35* | 60* | – | – | 0.2* | 80* | – | – |
| Si | 0.52 | 7.6 | 30 | 251 | 0.77 | 12* | 800 | 249 |
| Mn | 0.78 | 4.5* | 5.5 | 249 | 0.76 | 5* | 76 | 224 |
| P | 0.13 | 35 | 1 | 183 | 0.23 | 35 | 290 | 230 |
| S | 0.035 | 40 | 240 | 223 | 0.091 | 40 | 456 | 215 |
| Cr | 0.88 | 0.8 | 6.3 | 252 | 0.99 | 0.96* | 6 | 212 |
| Ni | 0.94 | 0.4 | 0.9 | 251 | 0.77 | 3** | 970 | 262 |
| Mo | 0.81 | 1.9 | 3.6 | 240 | 0.91 | 1.9 | 78.5 | 226 |

The cooling rate used was 250K/sec, which was measured under the same condition of the Trans-Varestraint test. The primary and the secondary dendrite arm spacings used were 20 and 6$\mu$m respectively which were measured in quenched weld metal by the linear intercept method. The division numbers $n_x$ ($=n_y$) used were 20 and 16 on the basis of the calculated BTR which will be shown in 3.1. Then $N_T$ is 400 and 256, respectively.

### 2.2.5 Temperature of solid bridge formation

Calculated relations between temperature and fraction solid $f_s$ in SUS310S and SUS430 are shown in Fig.6. The broken line means the relation in the case of parabolic growth. The lower limit of BTR means the temperature where solidification is completed. As mentioned in 2.2.1, the beginning of solid bridge is formed at $f_s$=0.5. Typical characteristics can be seen in SUS310S shown in (a). The temperature where the beginning of solid bridge is formed is 1666K in the model, namely only 7K beneath the solidification front. This corresponded

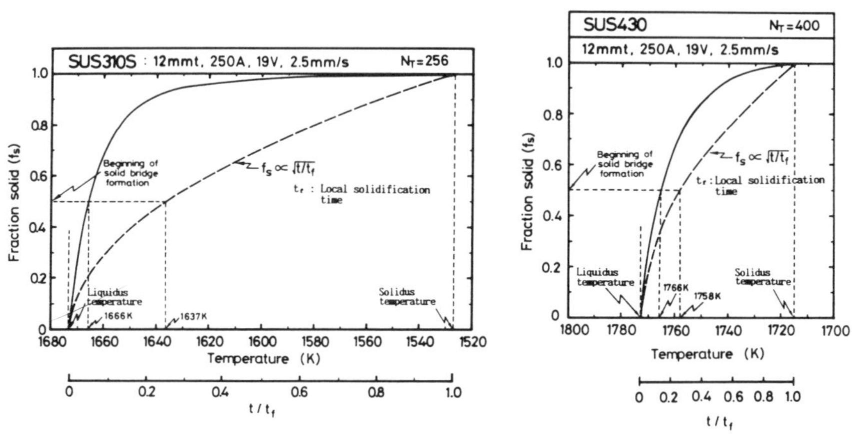


(a) SUS310S (b) SUS430

Fig.6 Change in fraction solid calculated with temperature drop during weld solidification

well to the observation by liquid-tin quenching, although the observed temperature seemed somewhat higher than the calculated ones. Therefore, the model means that small amount of liquid which is confined by the solid bridges remains for very long period in the BTR. For example, the temperature region where $f_s$ is larger than 0.9 covers about 1650 to 1527K. This tendency also corresponds well to the observation by liquid-tin quenching. By the way, Fig.7 shows the tendency of solute enrichment into the residual liquid in SUS310S, and means that P, S and Si are prone to concentrate. These are in good agreement with the experimental results which showed an evidence that traces of liquid droplets enriched with mainly P or S and somtimes with Si in a lower temperature region of BTR.

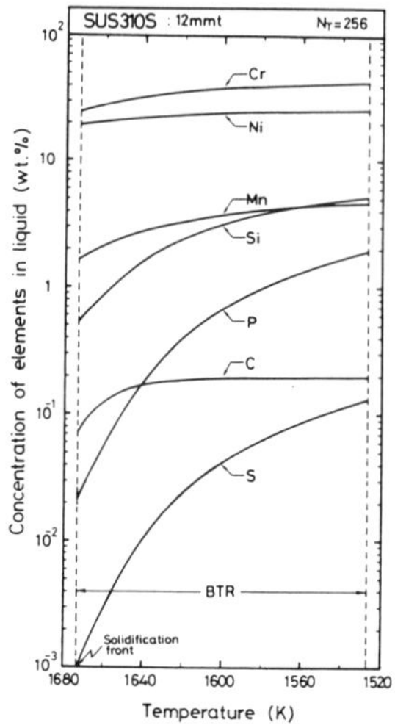


Fig.7 Change in concentration of elements in residual liquid during weld solidification of SUS310S

The behavior in SUS430 is shown in Fig.6(b). Fundamental behavior is the same as in (a). The temperature region between the temperature of solid bridge formation and the lower temperature limit of BTR is very narrow in SUS430. These behaviors corresponded to the observation by liquid-tin quenching.

2.2.6 Solute distribution in cellular dendrite

The solid curves in Fig.8 (a) and (b) show isoconcentntration curves of Ni and Cr just at the completion of solidification in SUS310S. The symbols O, H, P and W in the four corners are the same as in Fig.5, and the abscissa and the ordinate are normalized by $\overline{OW}$ and $\overline{OH}$, respectively. The numerals in parentheses attached to the broken lines mean Ni and Cr concentrations in solid at the moment where the solid-liquid interface was there during solidification. All of

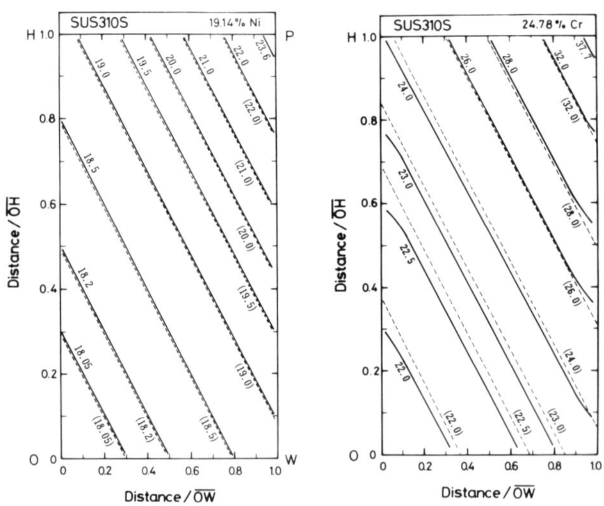


(a) Ni (b) Cr

Fig.8 Isoconcentration curves in cellular dendrite just after solidification of SUS310S

the solid curves almost correspond to the broken lines except somewhat bigger segregation of Cr, because of very slow diffusivity of Ni and Cr in austenite.

Comparison between the calculated and experimentally analyzed Ni and Cr concentrations are given in Fig.9 (a) and (b). The alphabetical letters used to denote the analyzed locations are defined in the upper right corner. Solid circles mean the experimentally analyzed concentration, and triangles mean the calculated concentration. Open circles mean calculated average concentration within a circle of 1$\mu$m radius considering the size of X-ray generating region in EPMA analysis. It should be noticed that the calculated results agree well with the experimental results except a small disagreement at the point P and O. The authors could explaind, however, that these disagreements were mainly caused by the value of equilibrium partition coefficient. For example, when $k_{Cr}$ is adopted 0.92 instead of 0.88 the disagreement was much improved.

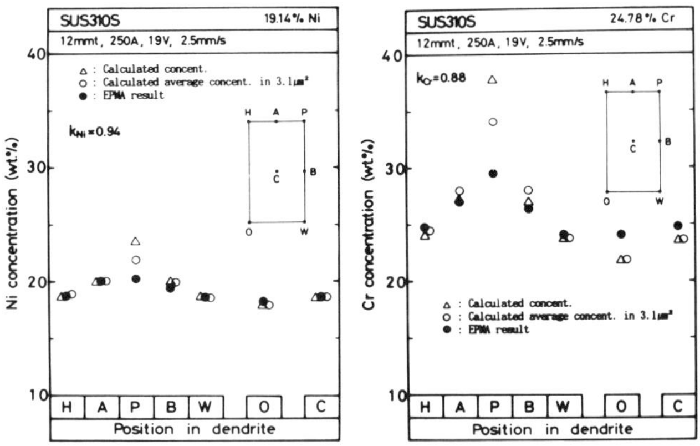


(a) Ni (b) Cr

Fig.9 Comparison between calculated and measured Ni and Cr concentrations in cellular dendrite of SUS310S

Figure 10 (a) shows isoconcentration curve of Ni just at the completion of solidification in SUS329J2L. The definitions of solid and broken curves are the same as in the results of SUS310S. In this case, the solid curves fairly deviate from the broken curves because of relatively faster diffusivity of Ni in ferrite. Especially near $\overline{OW}$, isoconcentration curve is nearly horizontal. This means that Ni is almost uniform along $\overline{OW}$ direction.

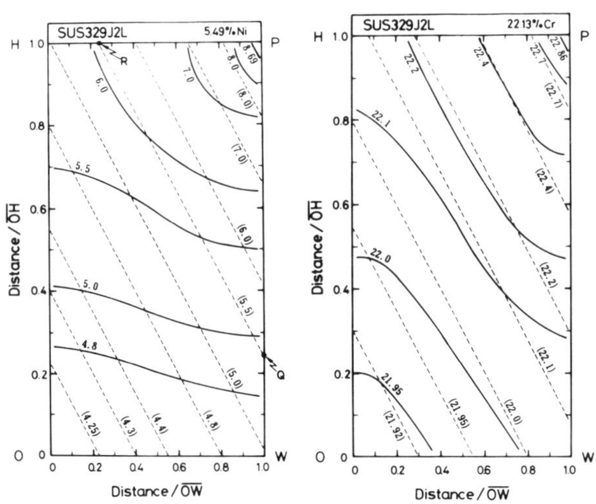


(a) Ni (b) Cr

Fig.10 Isoconcentration curves in cellular dendrite just after solidification of SUS329J2L

Comparison between the calculated and the experimentally analyzed Ni concentration is given in Fig.11 (a). A good correspondence between the calculated and the analyzed results can be seen. It should be noticed that the concentration at B is definitely lower than A, and that the concentration at W is also definitely lower than H. This tendency was not clearly seen in Fig.9 (a), and this is because of the faster diffusivity of Ni in ferrite. Similar difference in ferrite and austenite was seen in calculated distributions of S and P, and it might be explained by the well known fact that solute diffusivity in ferrite is generally faster than in austenite. The reason why the secondary dendrite arm boundary becomes obscure faster than the primary dendrite arm boundary in the solidification mode with primary ferrite may be explained by these behaviors. Figure 10 (b) and Fig.11 (b) show the behaviors of Cr in SUS329J2L, and a good correspondence between the calculated and the analyzed results can be seen.

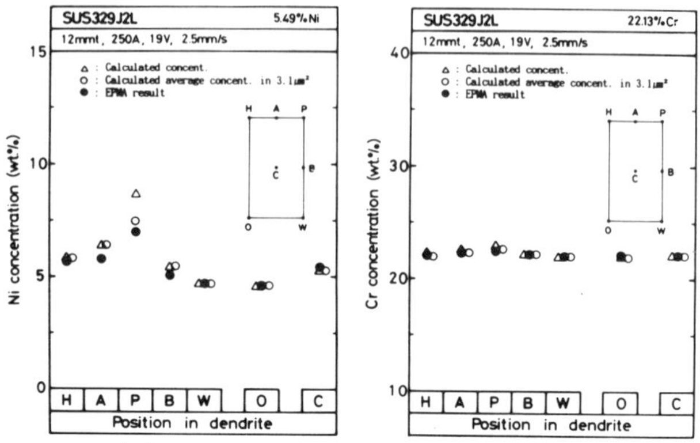


(a) Ni (b) Cr

Fig.11 Comparison between calculated and measured Ni and Cr concentrations in cellular dendrite of SUS329J2L

As well known, phosphorus and sulphur segregations are very important from the viewpoint of solidification cracking, but it is actually very difficult to analyze their distribution in dendrite because of their low content and low solubility. Then for information, Fig.12 (a) and (b) show phosphorus distribution in SUS310S and SUS329J2L, respectively. The segregation of phosphorus in SUS310S is severer than SUS329J2L because of smaller partition coefficient and slower diffusivity.

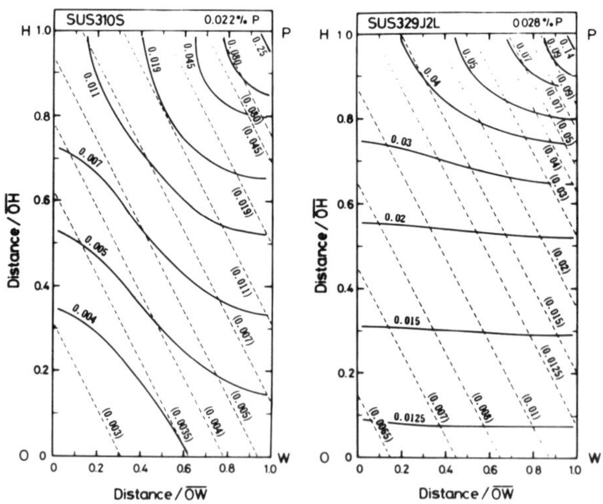


(a) SUS310S (b) SUS329J2L

Fig.12 Distributions of P in cellular dendrite just after solidification of SUS310S and SUS329J2L

## 3. Numerical analysis on Solidification Brittleness Temperature range (BTR) for crack susceptibility

### 3.1 Calculation of BTR and ranking of individual element for SUS310S and 430

The fractograph of solidification crack by the Trans-Varestraint test implied that the lower temperature limit of BTR corresponded to the temperature where small amount of residual liquid enriched with impurities solidified completely. Namely, it is guessed that the lower temperature limit of BTR has the meaning as the temperature where the last mesh solidifies in this model.

The calculation of BTR was determined as follows : In the modeling, the solid-liquid interface advances so that the liquidus temperature of the liquid may agree with the actual temperature. Impurities are enriched into residual liquid with the advance, and the temperature where the last mesh enriched with impurities solidifies gives the solidus temperature. Therefore, BTR was calculated as the difference between the initial liquidus temperature which was measured experimentally and the temperature where the last mesh solidifies.

In the modeling used (Fig.5), the division number ($n_x$, $n_y$) is important to calculate BTR, because the accuracy largely depends on the mesh size. As shown in Fig.2, solidification was finished by the formation of inclusion at the intersectional site of the primary and the secondary dendrite arm boundaries, which was enriched with P, S etc. according to EDX results. This implies that the residual liquid having nearly the same size as the inclusion was to be considered the smallest unit at moment when solidification was completed because of e.g. eutectic reation to form inclusion. Therefore, the authors decided the total number of division $N_T(n_x \times n_y)$ are 256 ($n_x = n_y = 16$) and 400 ($n_x = n_y = 20$), by the calculation of the inclusion size near the lower temperature limit of BTR. Matsuda et al (1977) has measured the BTR of many different types of SUS310S by changing P and S contents using the Trans-Varestraint test. These data and one datum of SUS430 were numerically analyzed here. The calculated and the experimental BTR are plotted in Fig.13, where only two data denoted by asterisk are the data of this time. The correspondence between the calculated and the experimental BTR is good inspite of widely changed P, S and Si contents and irrespective of primary solidification phase of austenite and ferrite.

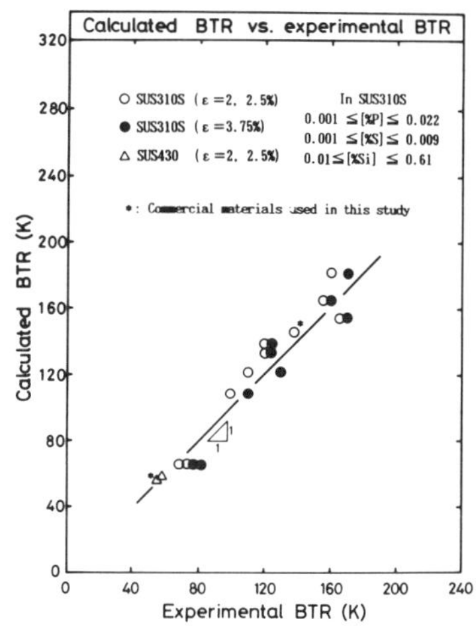


Fig.13 Comparison between experimental and calculated BTR

Meanwhile, BTR has the meaning as total depression of melting point from the initial liquidus temperature, and the total depression is expressed as the sum of the partial depression by each element. This means that the ranking of harmful elements is easily possible by comparing the ratio of partial depression to the total depression(BTR), namely the contribution of each element. The contribution of each element in commercial SUS310S is given in Fig.14 (a). Another example is shown in (b) where the P content is much lower, the S is higher and Si is lower than (a). From these two figures, it is seen that P and S and then Si are strongly contributory elements. This tendency has been well known through many experimental results.

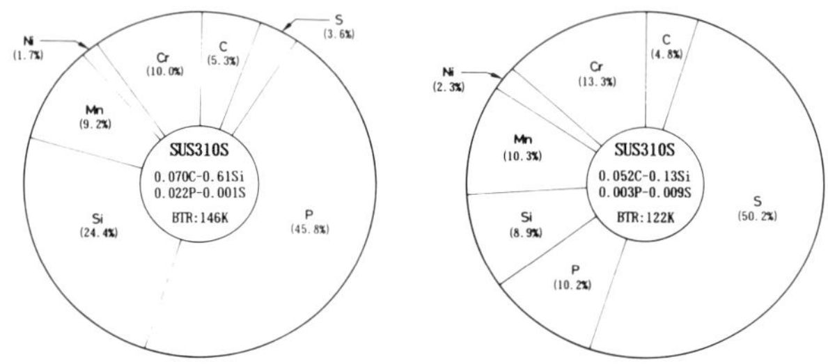


(a) Commercial material (b) Tentative material

Fig.14 Cotribution of elements to calculated BTR in SUS310S

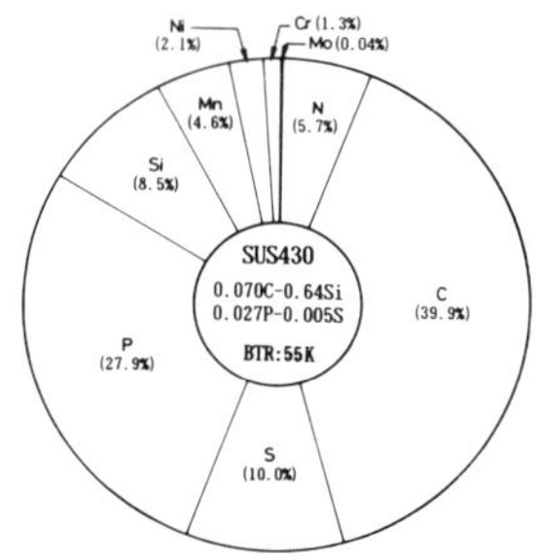


Fig.15 Contribution of elements to calculated BTR in SUS430

Figure 15 shows the contribution of elements in SUS430. In this result, the contribution of nitrogen was calculated on the same basis which is shown in 3.2. In this case, C is the most contributory, secondary P and then S. By the way, Kah et al(1981) reported that S, C, N and P are the primary cotributors to hot cracking in SUS430 by the Trans-Varestraint test. Although they derived this result from the measurement of total crack length, both of the calculated and their experimental results have a good agreement.

## 3.2 Calculation of BTR for duplex stainless steel

### 3.2.1 Prediction of harmful effect of nitrogen

The BTR of base metal and deposited metal of duplex stainless steel was calculated, and the results are listed in Table 2. It is seen that the calculated BTR was narrow compared with the experimental BTR. The authors considered that this discrepancy is due to the neglect of the harmful effect of nitrogen by the next reasons. Accoring to Fig.15 for SUS430 whose solidification mode is the same as duplex stainless steel, carbon has the biggest contribution to the BTR.

Table 2 Comparison between experimental and calculated BTR without taking account of N in SUS329J2L and its deposited metals

| Material | Experimental BTR (K) | Calculated BTR (K) |
|---|---|---|
| SUS329J2L(B) | 98 | 61 |
| DP70HN | 95 | 63 |
| DP80HN | 95 | 70 |
| DP100HN | 120 | 61 |

Considering that nitrogen is an austenite stabilizer as well as carbon concerning solidification mode, it can be estimated that nitrogen should be harmful, too. Moreover, nitrogen content in duplex stainless steel is usually high compared with general austenitic and ferritic stainless steel. Therefore, the authors tried to take account of nitrogen in BTR calculation. However, it is difficult to get the values of partition coefficient and liquidus gradient, because even new Fe-N phase diagram does not give any information around Fe corner.

With the above reason, the authors had to estimate the partition coefficient of nitrogen. Here the authors defined $C_{eq}$ as $C_{eq} = [\%C] + a_N[\%N]$, where $a_N$ is the coefficient to show effectiveness of nitrogen. From estimation of the microstructure at room temperature, the $a_N$ is predicted as 1.0 (after DeLong) or 0.65 (after Hammar). Therefore, the authors have tried to calculate using 1.0 and 0.65 as $a_N$. As a result of comparison between the experimental and the calculated ones the $a_N$ value of 1.0 was better than 0.65 in scattering of data. Therefore, in this study the effect of nitrogen was taken account of the carbon equivalent with $a_N=1$, which is expressed by the simple sum of [%C] and [%N]. According to this treatment, the contribution of nitrogen to the BTR is rather big as seen in Fig.16.

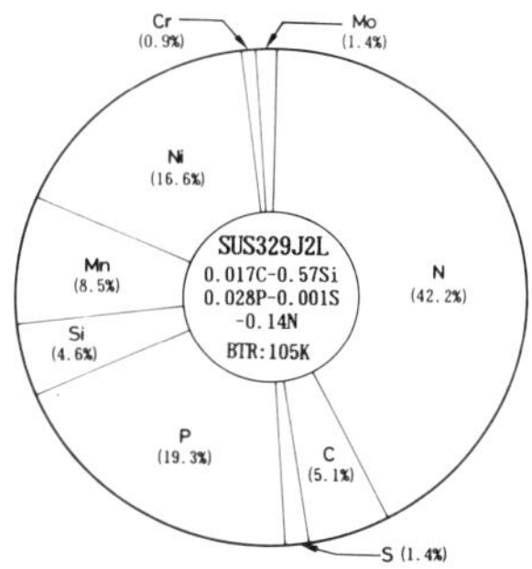


Fig.16 Contribution of elements to calculated BTR in SUS329J2L

3.2.2 Experimental confirmation of harmful effect of nitrogen

Tentative deposited weld metal whose nitrogen content was 0.02 to 0.14% and ferrite content was varied from about 0 to 100% was tested by the Trans-Varestraint test.

Figure 17 summarizes the maximum and the total crack length vs. ferrite content, in which the data were separated into two curves from nitrogen levels as 0.02-0.04% and 0.10-0.14%. In the range of ferrite content less than about few percentages, namely in the range of primary austenite solidification, the effect of nitrogen is not clear. This should be reasonable because nitrogen is austenite stabilizer. In the range of ferrite content higher than about 80%, a beneficial effect of reducing nitrogen content is clearly seen. It should be noticed that both the maximum and the total crack length of these low nitrogen deposited weld metal are almost the same

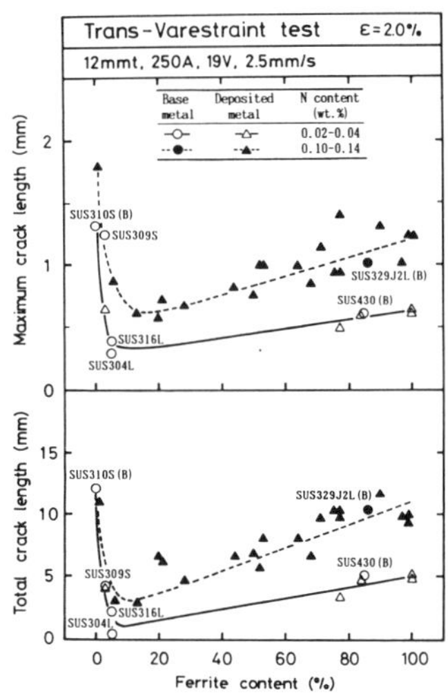


Fig.17 Maximum and total length vs. ferrite content under two levels of N content

as those of SUS430 base metal. Therefore, a harmful effect of nitrogen in the solidification mode in which ferrite is the primary phase was confirmed experimentally.

4. Conclusion

(1) Weld solidification sequence of various stainless steels which have different solidification modes as primary austenitic, ferritic and duplex phases was clearly observed by means of liquid-tin quenching. Consequently the following features were cleared : (a) Solid bridge was formed at the early stage of the solidification front by mutual contact of counter secondary dendrite arms. (b) Due to developing of the solid bridge, residual liquid was confined to intersectional site of the primary and the secondary dendrite arm boundary. (c) In weld metal solidified with primary ferrite, cellular dendrites were rapidly obscured with decreasing in temperature even within BTR. Moreover, the secondary dendrite arm boundary was preferentially obscured compared with the primary dendrite arm boundary, while this feature was not noticiable in weld metal solidified with primary austenite phase.

(2) A two dimensional modeling of cellular dendritic growth mode during weld solidification was made based on the solidification sequences of stainless steels which were revealed by liquid-tin quenching. Basic standpoints in the modeling are that solid-liquid interface advances so that the liquidus temperature of liquid may correspond to the actual temperature, and the solid bridge is formed as fraction solid is reached to 0.5.

(3) In this modeling, fraction solid increases very rapidly near liquidus temperature, and thus the beginning temperature where the bridge is formed is very close to the liquidus temperature. Therefore, the temperature zone where residual liquid is confined by the solid bridge is rather long. These behaviors were actually confirmed by liquid-tin quenching.

(4) Distributions and segregations of Ni, Cr and P in cellular dendrite during and after solidification were calculated in this modeling. Distributions of Ni and Cr calculated in cellular dendrite fairly correspond to the analysed results in the weld metals. Therefore, it results that the two dimensional modeling proposed in this paper is believed to be an useful model to estimate quantitatively the sequence of solidification and the formation of solute segregation.

(5) Using the modeling, Solidification Brittleness Temperature Range (BTR) which is evaluated as an important index for solidification crack susceptibility was estimated for various stainless steel base and weld metals. Moreover detrimental role of individual element in the BTR was evaluated. Also the results calculated were confirmed with the experiments. Consequently the following features were cleared : (a) The BTR calculated for fully austenitic and ferritic stainless

steels agreed well with the BTR experimented. It was possible to rank harmful elements by the calculation that the primary contributor to the BTR was P, S and then Si in fully austenitic and C, S and P in ferritic stainless steels. (b) Application of the modeling to duplex stainless steel showed that the BTR calculated agreed well with the BTR experimented if calculation is done with taking account of nitrogen content. (c) The harmful effect of nitrogen in the BTR which was calculated was confirmed experimentally with tentative deposited weld metals containing low and high levels of nitrogen. The contribution of nitrogen to the BTR was crucial in duplex stainless steel.

## References


Kou,S. and Le,Y. (1982) The Effect of Quenching on the Solidification Structure and Transformation Behavior of Stainless Steel Welds. Met. Trans., 13A-7, 1141-1152.

Matsuda,F. Nakagawa,H. and Lee,J.B. (1987) Weld Cracking in Duplex Stainless Steel (Report I). Trans. of JWRI, 16-2, 115-121.

Brody,H.D. and Flemings,M.C. (1966) Solute Redistribution in Dendritic Solidification. Trans. TMS-AIME, 236-5, 615-624.

Matsumiya,T. Kajioka,H. Mizoguchi,S. Ueshima,Y. and Esaka,H (1984) Mathematical Analysis of Segregation in Continuously-cast Slabs. Trans. ISIJ, 24, 873-882.

Ueshima,Y. Mizoguchi,S. Matsumiya,T. and Kajioka,H. (1986) Analysis of Solute Distribution in Dendrites of Carbon Steel with $\delta / \gamma$ Transformation during Solidification. Met. Trans., 17B, 845-859.

Ohnaka,I. (1986) Mathematical Analysis of Solute Redistribution during Solidification with Diffusion in Solid Phase. Trans. ISIJ, 26, 1045-1051.

McCartney,D.G. and Hunt,J.D. (1984) A Numerical Finite Difference Model of Steady State Cellular and Dendritic Growth. Met. Trans., 15A-6, 983-994.

Hunt,J.D. and McCartney,D.G. (1987) Numerical Finite Difference Model for Steady State Cellular Array Growth. Acta Metall., 35-1, 89-99.

Matsuda,F. Nakagawa,H. and Lee,J.B. (1989) Weld Cracking in Duplex Stainless Steel (Report II). Trans. of JWRI, 18-1, 107-117.

Matsuda,F. Nakagawa,H. and Lee,J.B. (1989) Weld Cracking in Duplex Stainless Steel (Report III). Trans. of JWRI, 18-1, 119-126.

Matsuda,F. Arata,Y. and Katayama,S. (1977) Solidification Crack Susceptibility in Weld Metals of Fully Austenitic Stainless Steels (Report II). Trans. of JWRI, 6-1, 105-116.

Matsuda,F. Katayama,S. and Arata,Y. (1981) Solidification Crack Susceptibility in Weld Metals of Fully Austenitic Stainless Steels (Report V). Trans. of JWRI, 10-2, 201-212.

Kah,D.H. and Dickinson,D.W. (1891) Weldability of Ferritic Stainless Steels. Welding J., 60-8, 135s-142s.

# MATERIAL ASPECTS OF CERAMIC-CERAMIC AND CERAMIC-METAL BONDING

M.G. NICHOLAS
Materials Development Division, Harwell Laboratory.

Abstract
The projected increased use of ceramics in advanced engineering frequently depends on the successful achievement of ceramic-ceramic or ceramic-metal bonding. A range of joining techniques could be used to achieve bonding, but it is argued that those most relevant to structural applications are diffusion bonding using interlayers, brazing and glazing. Material aspects of these and some other techniques are reviewed and suggestions are made about future developments that could enhance the use of ceramics.
Key words: Bonding, Diffusion bonding, Brazing, Glazing, Ceramic, Ceramic-metal.

## 1. Introduction

The ability to produce reliable ceramic-ceramic and ceramic-metal joints is a key enabling technology for many advanced engineering projects. Design and performance demands are now being made that cannot be met readily by commercially developed metals or alloys, and considerable attention is being paid to the potential usefulness of ceramics as structural materials in applications such as aircraft engines. Ceramics are generally more refractory and less dense than metals, Palm (1984), Smithells (1976) as illustrated by the data assembled in Table 1 and can be stronger. However, if these properties are to be exploited it is generally essential for an adequate joining technology to be available. As a rule, ceramics cannot be formed readily into complex shapes and hence it can be necessary or at least convenient to construct ceramic engineering artefacts by joining geometrically simple components. Similarly, the first applications of ceramics are often as inserts in otherwise metallic structures to which they must be secured strongly, and sometimes also hermetically.

In principle a very wide spectrum of techniques could be employed, but in practice the options available are restricted because of the influence of basic physical and chemical characteristics that present challenges to both designers and materials scientists. Thus the brittleness of ceramics makes it important for designers to use joint configurations that accommodate mismatches in the thermal contraction and elastic characteristics of the workpiece being joined. The

Table 1. Some typical ceramic and metal properties.

| Material | Melting Temperature °C | Density $kg.m^{-3}$ | Expansivity $K^{-1} \times 10^6$ | Thermal Conductivity $W.m^{-1}K^{-1}$ |
|---|---|---|---|---|
| $Al_2O_3$ | 2050 | 4000 | 8.0 | 33 |
| SiC | 2700$^{(2)}$ | 3150 | 4.5 | 90 |
| $Si_3N_4$ | 1900 | 3200 | 3.1 | 18 |
| $ZrO_2$$^{(1)}$ | 2700 | 5700 | 9.5 | 2.5 |
| Al | 660 | 2700 | 23.5 | 23.8 |
| Fe | 1536 | 7870 | 12.1 | 78.2 |
| Nb | 2467 | 8600 | 7.2 | 54.1 |
| Ni | 1455 | 8900 | 13.3 | 88.5 |

(1) CaO stabilised.
(2) Sublimes.

materials scientist, however, must overcome chemical characteristics that militate against successful bonding through surface treatment of the workpieces and careful selection of the bonding agents.

Difficulties in joining ceramics using metal oriented technology have been recognised for a long time, and considerable effort has been put into the development of adequate joining technologies for a number of established ceramics, such as the molybdenum-manganese coating of siliceous alumina prior to its brazing. However, the number of ceramics of engineering interest is increasing and development work on joining technologies for these materials is intense at present. This paper will attempt to summarise the current situation and suggest some future developments. Attention will be concentrated on processes relevant to structural applications where high temperatures may be encountered in service conditions, and electronic applications will be ignored deliberately.

## 2. Direct bonding techniques

### 2.1 Fusion welding

Fusion welding is used to join glasses when making graded seals but is rarely employed to form non-vitreous ceramic-ceramic joints. The principal reason for this rarity is probably the high melting temperature of industrially important ceramics but fusion welding ceramics to metals poses even more challenges due to the generally large differences in melting temperature, thermal conductivity and contraction characteristics of the workpieces - and the uncertain chemistry and properties of the weld pool. Nevertheless fusion welding has been used in laboratories to produce ceramic-metal joints.

Electron beam welding was used to make prototype electrical feed throughs of Kovar, molybdenum, niobium or tungsten embedded in alumina insulators, Hokasson (1963). Successful fusion welding of molybdenum to alumina was described also by Pattee (1968), while Rice (1970) described reporting a major study of the feasibility of electron beam and arc welding alumina to tantalum, and zirconia to molybdenum, niobium or tantalum. More recently, More (1985) has described an attempt to fusion weld silicon carbide which failed because of its tendency to sublime rather than melt and form a stable weld pool.

While of interest and possible, fusion welding is unlikely to be of practical utility for any but a few exceptional ceramics and ceramic-metal systems and therefore the major attention in this survey will be focussed on other techniques. Of these, direct diffusion bonding between the workpieces most resembles fusion welding.

## 2.2 Diffusion bonding

The diffusion bonding of some metal workpiece materials is a well established practice with joints being fabricated by the application of compressive loads insufficient to cause macroscopic plastic deformation times ranging from minutes to hours at temperatures typically ranging from two-thirds to nine-tenths of the workpiece melting temperature, 0.67-0.9 Tm, K.

Because of the lack of deformation it is usual to promote contact between the workpieces by making their mating surface very flat and smooth. Sophisticated models have been developed that make quantitative predictions of the effects of temperature, time, surface texture and other parameters on the extent of bonding of metals and some of these models can be applied also to the diffusion bonding of ceramic-ceramic systems. The experimental data neeeded to test their validity are sparse but there is some evidence of similarity to metal-metal phenomena for both ionic and covalent ceramics. Thus alumina has been diffusion bonded by application of low compressive loads at 0.72 Tm followed by pressureless heating at 0.92, Scott (1985), and silicon carbide workpieces have been bonded at 0.65 and 0.75 of the sublimation temperature, 3000K, using fine powders introduced between the workpieces, More (1985) and Iseki (1980). While adequate strengths have been achieved, Scott (1985), it is difficult to create a ceramic-ceramic bond because of the need to match the spatial charge distributions at interfaces and it is noteworthy that the presence of sintering aids such as magnesia in alumina, Scott (1985), can be beneficial.

A more common practice than ceramic-ceramic bonding is ceramic-metal bonding. Unless special care is taken in the selection of the ceramic and metal workpiece materials, their interfaces can be degraded by marked thermal contraction mismatch stresses. However, one system in which these can be successfully accommodated is alumina-niobium and considerable experimental effort has been devoted to developing its diffusion bonding as a fabrication process for high pressure sodium lamps. The diffusion bonding of alumina-niobium has been used also as a model system and laboratory studies have evaluated effects of bonding temperature, pressure, time and environment as well as surface finish, orientation and grain size, Morizumi (1981) and Elssner (1989). Optimum conditions yield strong tough joints with fracture energies of 60 J $m^{-2}$, Elssner (1989), and high degrees of lattice continuity, Rühle (1986).

The diffusion bonding of other ceramic and metal workpieces has not been as thoroughly studied, perhaps because it is a rare industrial practice. Much more commonly, ceramic-ceramic and ceramic-metal systems are diffusion bonded indirectly by employing soft and ductile metal interlayers.

## 3. Indirect bonding techniques

### 3.1 Diffusion bonding using interlayers

It is usual to use high compressive loads when diffusion bonding indirectly, forcing the soft metal interlayer to flow macroscopically and conform to the surface topographies of the relatively rigid ceramic or metal workpieces. Thus it is not necessary to prepare very flat smooth workpiece surfaces and the bonding temperature employed can be low if the interlayer Tm is low, so avoiding detrimental microstructural coarsening of the workpieces and diminishing mismatched contraction strains.

The indirect technique of diffusion bonding has found application in high technology applications such as the fabrication of ceramic electrolyte batteries, observation ports for fusion experiments, and high energy accelerator modules, Figure 1, Joy (1987). The most commonly bonded ceramic in the workshop and in the laboratory is alumina . Thus at Harwell we have used alumina to produce accelerator prototypes and have used ASTM test pieces to evaluate the effects of fabrication pressure, temperature and time, and of aluminium foil thickness on the strengths of alumina/aluminium/steel joints, Nicholas

Fig. 1 220 mm diameter titanium-alumina accelerator module bonded using aluminium interlayers. Reproduced by courtesy of Dr. T. Joy, UK Science and Engineering Research Council, Daresbury Laboratory.

(1988). While Figure 2 shows that all the process parameters had an effect, that of the bonding temperature was most dramatic, the dependence being describable by an equation

$$B.S. = Bo.e^{-Q/RT}$$

where Bo is 3.14 x $10^6$ MPa and Q is 83.3 $kJ.mol^{-1}$. The value of the apparent activation energy Q has been compared by Derby (1987) to that for a number of potentially relevant processes and he associated it with the 84 $kJ.mol^{-1}$ for grain boundary diffusion of aluminium.

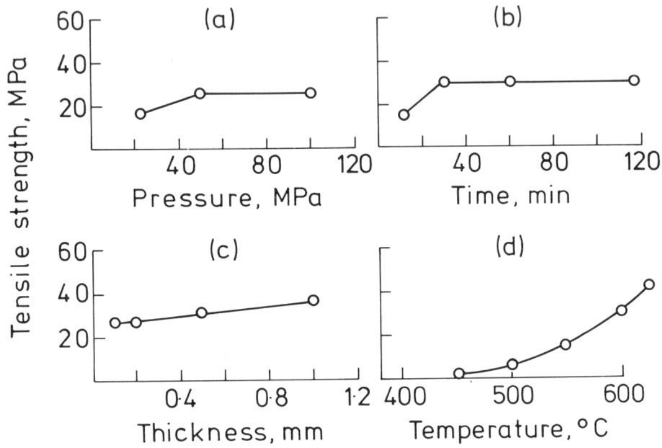


Fig. 2 The effects of varying some fabrication parameters on the tensile strenths of alumina-stainless steel samples diffusion bonded using aluminium foil interlayers; (a) pressure, (b) time, (c) interlayer thickness, (d) temperature. Standard conditins: 50 MPa applied in vacuum for 30 minutes at 595°C using a 0.5 mm thick interlayer.

Similar temperature dependencies have been observed when metal foils are used to form alumina-alumina joints, Figure 3, with the Q for aluminium being 80.4 $kJ.mol^{-1}$. Apart from the steep temperature dependence of the bond strength values it is noteworthy that there is an effective initiation temperature of 0.66-0.72 Tm for all four metals and that copper and nickel and possibly silver produce bond strength peaks at well defined temperatures. A similar peak in the fracture energy and toughness of alumina-copper bonds has been reported recently by Elssner (1989) - albeit at a somewhat higher temperature as also was that for the initiation of bonding. However peaks are not specific to alumina or even oxides, being observed for diffusion bonding diamonds to copper, Dewar (1976).

While frequently observed, the causes of these peaks are not always obvious. In chemically inert systems they could reflect the increasing severity of mismatched contraction stesses as the bonding temperature is raised. Similarly in reactive systems the ultimate degradation in bond strength could be due to the growth of thick fragile reaction product layers. These could be produced by direct interaction of the interlayers with the workpieces - or to the influence of the bonding environments. Thus recent work has demonstrated that use of ultra high rather than technical vacuum environments can decrease the initiation temperature for bonding, Pepper (1976) and produce much tougher joints, Elssner (1989). On the other hand, there is also evidence that joints formed in mildly oxidising technical vacua can be stronger than those produced in reducing environments, Nicholas (1987) and Calow (1971).

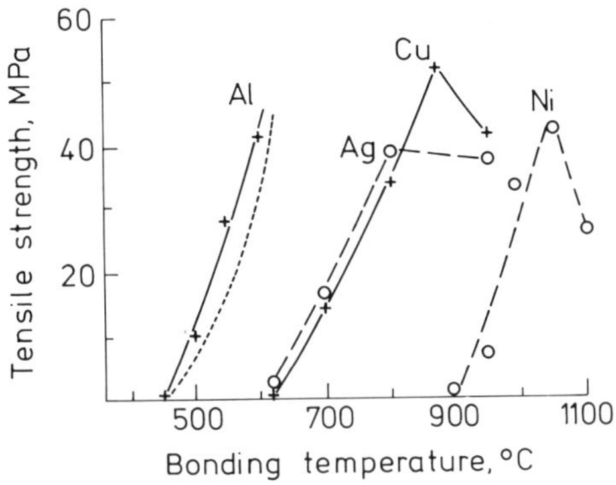


Fig. 3 The strengths of joints between alumina ASTM tensile test pieces bonded using 0.5 mm thick interlayers. A pressure of 50 MPa was applied in each case for 30 minutes in vacuum. The dotted line refers to Figure 2 data.

3.2 Brazing

The first requirement for a braze is that it wets the workpieces, spreading over surfaces and entering capillary gaps between components to be bonded. However, conventional brazes such as the silver-copper eutectic wet few ceramics of technical interest, Naidich (1981), basically because of the incompatible lattice binding characteristics of metals and ceramics. If ceramics are to be wetted, their surface chemistries must be modified by reaction with the brazes or by coating.

This necessity has led to the development of reactive braze alloys employing titanium as the reactant, although many other additives are suggested in the patent literature. Many factors may contribute to the commercial dominance of titanium but a scientific basis for it is suggested by simple analyses of braze-ceramic reactions using thermodynamic data.

Titanium can form a family of oxides, Massalski (1986), but thermodynamic analyses predict that reactions with alumina will produce TiO at 1400K (1127°C) if the titanium activity exceeds $4 \times 10^{-4}$, Nicholas (1987). At lower activities, $Ti_2O_3$ and other titanium lean oxides will be formed and this variation in the reaction produce chemistry is important because TiO differs by exhibiting metallic characteristics, Rostoker (1952), and being wettable by copper, Naidich (1981). Thus copper alloyed with sufficient titanium to have an activity exceeding $4 \times 10^{-4}$ should wet alumina by forming TiO - and this has been observed in practice, Naidich (1974). If a solvent other than copper is used, the critical titanium activity is unaltered but the titanium concentration will differ, generally being lowest for those solvents in which titanium is least soluble. For non-oxide ceramics such as carbon, similar effects are observed, Standing (1978), with wetting induced by high titanium activities being associated with the formation of hypostoichiometric TiC, Nicholas (1987).

Most active braze alloys have ternary or more complex compositions, and this can be an advantage if the solubility of titanium in these more complicated solvents is low. Thus while adding titanium to copper promotes its ability to wet, it also increases the severity of the reaction and can lead to the formation of fragile thick reaction product layers. Ideally, the titanium should be virtually insoluble so that little is needed to induce wetting and the reaction product layers are thin. Titanium is less soluble in silver than copper, and this suggests one reason why commercial active braze alloys are usually based on the silver-copper eutectic rather than copper. If metals such as tin or indium are added to copper they decrease the titanium solubility even more than does silver and allow alumina to be wetted at much lower titanium concentrations, Figure 4, Valentine (1980). While the figure refers to laboratory alloys, it is noteworthy that commercial silver-copper-indium-titanium brazes have been developed with very low levels of titanium, Mizuhara (1985).

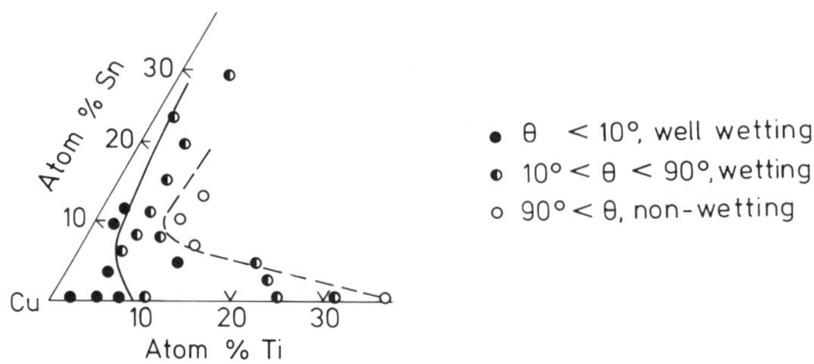


Fig. 4 The wettability of alumina by copper-tin-titanium alloys. θ is the sessile drop contact angle.

Once wetting is achieved it is necessary for strong bonds to be produced if useful joints are to be formed. As with diffusion bonding, the growth of reaction product layers can promote or degrade bonding, Scott (1975). For binary reactive alloys, the peak strengths achieved with alumina can be related to the affinity of the reacting additive for oxygen, Nicholas (1987). In practice, factors such as the topography and mechanical integrity of the ceramic surface can have major effects on strength values, Mizuhara (1985).

Active metal brazes are usually stiff, but it is advantageous to use ductile brazes if thermal expansion mismatches have to be accommodated. Thus aluminium is frequently used to braze low expansion silicon nitride ceramics, Naka (1986) and Ning (1987), and aluminium-silicon alloy has been used to braze a number of silicon ceramics, Yamada (1987), the relative ductility of braze permitting quite marked mismatches of contraction characteristics to be accommodated to some degree, Figure 5.

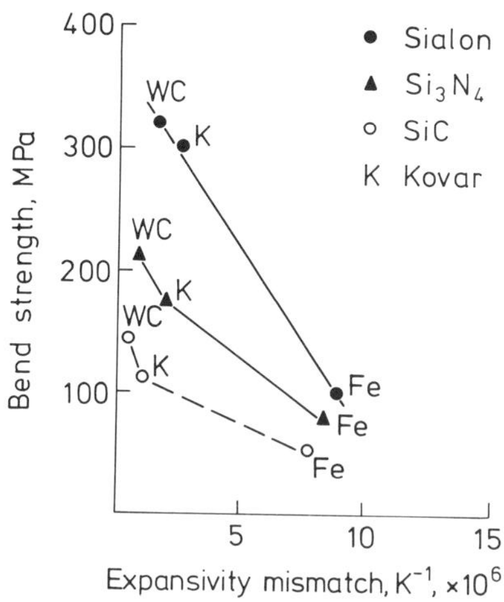


Fig. 5 Room temperature bend strengths of dissimilar workpieces brazed with aluminium - 10% silicon, after Yamada (1987).

The other approach, and the major brazing technique for ceramics, is to metallise their surfaces prior to joining with ductile conventional brazes. The most widely used method of metallising ceramics is the moly-manganese process in which a mixture of molybdenum or $MoO_3$ and manganese or $MnO_2$ is applied to the surface of a silicon bonded oxide ceramic and fired in damp hydrogen at about 1400°C to produce a coating of molybdenum particles held in a glass frit that is continuous with the ceramic binder phase, Twentyman (1975). This technique is very well established and variants have been developed that can be used with binder free ceramics. However, considerable skill is required if unacceptably high reject rates are to be avoided and this has led many fabricators to consider direct application of metal coatings by processes such as chemical vapour deposition or sputter ion plating. These produce thin overlayers and hence care has to be taken to avoid their dissolution in the braze and subsequent dewetting, van Thoor (1989).

3.3 Glazing

Many glasses have been developed as bonding agents for ceramic-ceramic and ceramic-metal joints. Glasses are oxides and therefore it is not surprising that they wet oxide ceramics very well forming interfaces with continuity of anion-cation bonding, Pask (1971). However, if metals are to be wetted it is advantageous to oxidise their surfaces choosing conditions that yield adherent films of a thickness that does not dissolve rapidly in the glass. Thus glazing metals and brazing metallised ceramics are mirror images.

Achieving satisfactory wetting of non-oxide ceramics can pose problems, but adequate techniques are being developed. Thus the ceramic could be coated with a metal whose outer surfaces was then oxidised. Another approach is based on the presence in commercial ceramics of glassy binder phases that can be exploited as glazes. Johnson (1985) used $SiO_2$-MgO-$Al_2O_3$, $Y_2O_3$-$SiO_2$-$Al_2O_3$ and $Y_2O_3$-$SiO_2$-$Al_2O_3$-N glasses to bond commercial quality silicon nitride, doped with magnesia or yttria, by heating in a nitrogen environment. Again, Iwamoto (1986) used reactive CaO-$TiO_2$-$SiO_2$ glasses to bond silicon nitride in a nitrogen environment, forming a layer of TiN at the interface. Somewhat similar to such reactive glazing is the use of non-vitreous oxide frits that bond by forming eutectic melts with oxide ceramics, Swatz (1989) and Jonker (1963). Also related to this variant of glazing is the use of non-oxide frits, such as the molten $SiB_6$ and $MoSi_2$ used to bond silicon carbide, More (1985).

While wetting permits a bond to form, it is vital that it remains intact as the joint cools. Glasses do not have clear melting temperatures but become increasingly viscous as they cool. Nevertheless temperatures can be identified at which glasses flow and become rigid or set, Holloway (1973). As the glass cools it shrinks, but in general the coefficients of expansion are low with the most refractory glass being the least expansive, Shand (1958). Because of the inherent brittleness of glasses, it is usually important that the contraction characteristics of the rigid workpieces and the glaze should be well matched if fracture is not to occur as bonded components cool. In general, joints will remain intact if the contractions differ by less than 0.01% and as much as 0.1% may be tolerated by small assemblies.

A few glass compositions have been developed as glazes for high expansion workpieces, but these are generally not be very refractory and hence unsuitable for use with components that must endure high temperature service conditions. However in recent years the range of potentially useful glazes has been extended by the development of glass-ceramics, McMillan (1979). These materials are glasses that devitrify, crystallise, as they cool, and their compositions can be tailored to match those of many ceramic and metal workpieces.

### 3.4 Adhesive bonding

Adhesive bonding is widely used for the joining of both ceramics and metals for low temperature applications. While adhesives are still being developed, it is unlikely that even polyimide and bismaleimide adhesives will be able to operate in excess of 300°C and therefore they will be unsuitable for many applications of engineering ceramics such as silicon carbide and silicon nitride.

While much less refractory than brazes or glazes, the use of adhesives exhibits some phenomological similarities to brazing and glazing. Thus the workpiece surfaces must be cleaned carefully before bonding, and metal workpieces should have firmly adherent oxide surface films so that chemical bridges can be formed such as the -Fe-O-Si≡ sequence between steel and polysiloxanes, Gettings (1977). Very specific and detailed procedures have been developed for the preparation of metal workpieces, but technique development for ceramics has not been extensive and hence many bonding procedures are still largely based on experience and craft concepts rather than scientific understanding or validated engineering practice.

## 4. Future developments

Many techniques have been developed or evaluated for bonding ceramic-ceramic or ceramic-metal systems to satisfying existing industrial needs. However, this is a time of rapidly changing material applications and the next decade will see many new developments. Further, such are the timescales of advanced technology, such as the introduction of ceramics into aircraft engines, that joining techniques will have to be defined as the materials are being specified rather than afterwards. It is unlikely that any new classes of workpiece materials will be adopted as engineering ceramics during the next decade, Lewis (1989), but structural refinements could pose problems to the joining technologist. This could be particularly true for non-oxide ceramics subject to dissociative or oxidative degradation whose practical use may require doping with stabilising additives or coating with thermodynamically stable systems. These developments will place an immediate premium on knowledge based joining processes derived from generic studies that can accommodate workpiece modifications with confidence. Thus the next few years should see both developments relating to specific joining processes and generic studies to provide better understanding of interface characteristics and of joint performance.

I doubt whether the next decade will see fusion welding or direct diffusion bonding establishing themselves as major joining technologies, and it is unlikely that adhesive bonding will make a significant contribution to the application of high temperature ceramics. However, it is probable that there will be considerable development of indirect diffusion bonding, brazing and glazing.

Developments of indirect diffusion bonding could include the wider use of hot isostatic pressing techniques to increase the range of geometries that can be bonded and also through the use of large H.I.P. facilities to increase the throughput per unit time and make indirect diffusion bonding more accessible as a mass production process. The quality of the bonding achieved could be enhanced by deliberately encouraged interlayer-ceramic reactions, achieved by cladding the metal interlayers with reactive species or by varying the chemical activity of the bonding environments.

The use of active braze alloys will increase through the mechanism of market forces and through technical developments - in particular the use of active braze alloys as claddings on inserts that help to accommodate mismatched workpiece contraction characteristics and the emergence of new braze families. These new brazes should include some based on nickel for high temperature service, and their development as thin foils could be aided by application of rapid solidification techniques. In these developments of brazing and diffusion bonding thermodynamic analyses and detailed microstructural evaluations will play important roles in providing the needed scientific understanding.

Glazes and associated bonding agents should also experience vigorous development, particularly for high temperature applications of ceramic-ceramic systems. The glazes employed, however, are less likely to be glass sealants than vitreous binder phases naturally occurring in commercial ceramics. Similarly, high temperature requirements could lead to a wider interest in inorganic bonding agents used either as frits or produced as reaction products from diffusion bonded metal interlayers or coatings.

These developments will assist the fabricator, but it is performance that is of greater interest to the user. Designers need engineering data in a form that they can use, and providing this information will require extensive experimental measurement of joint characteristics and of modelling to provide a basis for interpretation. This work will be made easier if a number of joint configurations and testing techniques can be adopted as standards or normal practice. This need is widely recognised but little concerted action has yet to be taken to satisfy it - this therefore should be a major development in the next few years.

When such information becomes available it should be possible to codify the knowledge base as expert system as has been done already for the adhesive bonding of metals. However, I doubt whether it is feasible to use such systems to turn novices into experts since they will contain both explicit and implicit value judgements. Nevertheless, the systems could be invaluable knowledge stores when accessed by experts, and in general I believe the next few years will see a rapid increase in the quantification and codifying of joining technology for ceramics.

## References

Calow, C.A., Bayer, P.D. and Porter, I.T. (1971) J. Mat. Sci., 6, p. 150.

Derby, B. (1987) Mat. Sci. Res., 21, p. 319.

Dewar, B., Nicholas, M.G. and Scott, P.M. (1976) J. Mat. Sci., 11, p. 1083.

Elssner, G. (1989) in Designing Interfaces for Technological Applications (ed. S.D. Peteves), Elsevier Applied Science, London, p. 77.

Gettings, M. and Kinloch, A.J. (1977) J. Mat. Sci., 12, p. 2511.

Hokasson, A., Rogers, S.L. and Kern, W. (1963) Ceram. Ind. J., 81, p. 44.

Holloway, D.G. (1973) Physical Properties of Glass, Wykeham.

Iseki, T., Arakawa, K. and Suzuki, H. (1980) J. Mat. Sci., 15, p. 1049.

Iwamoto, N., Umesaki, N., Haibora, Y. and Sibuya, K. (1986) in Ceramic Materials and Components for Engines (ed. W. Bunk and H. Hausner), D.K.G., Bad Honef, p. 467.

Johnson, S.M., Rowcliffe, D.J., Brittain, R.D. and Lamoreaux, R.H. (1985) Ber. D.K.G., 1, p. 109.

Jonker, G.H., Klomp, J. and Botden, T.P.J. (1963) Sci. Ceram., 2, p. 295.

Joy, T., Crispin, R.M. and Nicholas, M.G. (1987) in High Technology Joining, B.A.B.S., Abington, p. 23/1.

Lewis, M.H. (1989) in Designing Interfaces for Technological Applications (ed. S.D. Peteves), Elsevier Applied Science, London, p. 271.

Massalski, T.B. (1986) Binary Alloy Diagrams, A.S.M., Menlo Park.

McMillan, P.W. (1979) Glass Ceramic Materials, Academic Press.
Mizuhara, H. and Huebel, E. (1985) Weld. J., 65, p. 27.
More, T.J. (1985) J. Am. Ceram. Soc., 68, p. C151.
Morozumi, S. (1981) J. Mat. Sci., 16, p. 2137.
Naidich, Y. (1981) Prog. Surf. and Membrane Sci., 14, p. 354.
Naidich, Y. and Zhuralev, V.S. (1974) Refractories, 14, p. 55.
Naka, M., Mori, H., Kubo, M., Okamoto, I. and Fujita, H. (1986) J. Mat. Sci. Lett., 5, p. 696.
Nicholas, M.G. (1987) Studies in Phys and Theoret. Chem., 48, p. 25.
Nicholas, M.G. and Crispin, R.M. (1988) in Diffusion Bonding (ed. R. Pearce), Cranfield Inst. Tech., Cranfield, p. 173.
Ning, X.D., Suganuma, K., Morita, M. and Okamoto, T. (1987) Phil. Mag. Lett., 55, p. 93.
Palm, B. (1984) in Ceramics in Advanced Energy Technologies (ed. H. Kröckel, E. Merz and O. Van der Biest), Reidel Publ., Dordrecht, p. 178.
Pask, J.A. (1971) Proc. Porcelain Enamel Inst. Tech. Forum, 33, p. 1.
Pattee, H.E., Evans, R.M. and Munroe, R.E. (1968) N.A.S.A. S.P.5052.
Pepper, S.V. (1976) J. App. Phys., 47, p. 801.
Rice, R.W. (1970) N.R.L. Rept. 7085.
Rostocker, W. (1952) Trans. A.I.M.E., 194, p. 981.
Rühle, M., Backhouse-Riccoult, M., Burger, K. and Mader, W. (1986) Mat. Sci. Res., 21, p. 395.
Scott, C. and Tran, V.B. (1985) Ceram. Bull., 64, p. 1129.
Scott, P.M., Nicholas, M.G. and Dewar, B. (1975) J. Mat. Sci., 10, p. 1833.
Shand, E.B. (1958) Glass Engineering Handbook, McGraw-Hill, New York.
Smithells, C.J. (1976) Metals Reference Book, Butterworths, London.
Standing, R. and Nicholas, M.G. (1978) J. Mat. Sci., 13, p. 1509.
Swatz, S.L., Majunder, B.S., Skidmore, A. and Mutsuddy, B.C. (1989) Mat. Lett., 7, p. 407.
Twentyman, M.E. (1975) J. Mat. Sci., 10, p. 765.
Valentine, T.M., Nicholas, M.G. and Waite, M.J. (1980) J. Mat. Sci., 15, p. 2197.
van Thoor, C.M.E., Hensworth, J.A., Nicholas, M.G. and Crispin, R.M. (1989) Effects of Titanium on some Surface Properties of Nitride Ceramics, AERE-R3566; also J. Mat. Sci., In press.
Yamada, T., Kohno, A., Yokoi, K. and Okada, S. (1987) Studies in Phys. and Theoret. Chem., 48, p. 489.

PART FOUR

# Research Directions in Welded Construction

# WELDING METALLURGY AND PROCESS DEVELOPMENT RESEARCH FOR THE GAS PIPELINE INDUSTRY

A.B. ROTHWELL, D.V. DORLING AND A.G. GLOVER
NOVA Corporation of Alberta

Abstract
The construction of gas transmission pipelines in Canada, as in other areas of the world, has placed heavy reliance on the use of traditional, manual SMAW techniques for field joining. Hydrogen-assisted cold cracking has been the most significant problem area with this process and, perhaps surprisingly, related research is still being conducted. The pursuit of greater productivity and reproducibility has led to wider use of mechanized GMAW; the limitations inherent in the original, short-circuiting process have led to both metallurgical investigations aimed at improving toughness and, more recently, process developments intended to bring metal transfer and metallurgical reactions under closer control. Current high levels of activity in the industry have also necessitated improvements in efficiency for small diameter pipelines and piping systems, and a number of developments, using both traditional and innovative welding processes, are in progress. Increasing application of a fitness-for-purpose approach to weld acceptance has led to a reassessment of NDE techniques, and to the development of mechanized ultrasonic testing techniques.
Key words: Pipeline, Arc-welding, SMAW, Pulsed GMAW, MIAB, Cracking, Toughness, NDE.

## 1. Introduction

The construction of gas transmission pipelines in Canada has made extensive use of the traditional, manual SMAW process, using cellulosic electrodes and downhill progression. The high hydrogen content and low heat input associated with this process, together with the low temperatures commonly encountered, make hydrogen-assisted cold cracking a significant issue. A good deal of research has been directed towards this problem over the years and, perhaps surprisingly, for such a mature technology, investigations are still continuing. Some on-going research issues in this area will be discussed in detail in the following section of this paper.

Cross-country pipelining involves repeating the same girth weld from forty to eighty times per kilometre. It is evident that such an operation should readily lend itself to mechanization, despite the potential problems posed by the field environment. Mechanized GMAW was first used by NOVA almost twenty years ago; until 1989, the necessary all-position capability had been achieved exclusively by the use of the short-circuiting mode, with its attendant limitations on consumables and control of metal transfer. Recent research has aimed at the development of pulsed GMAW techniques for pipeline welding; this has involved the use of flexible, programmable power sources for the optimization of pulse structure for the specific application envisaged, together with the exploitation of improved shielding gases which this process allows. The results of this work, which led to the first field application in 1989, are summarized below.

One of the most characteristic features of recent developments, for the Canadian gas transmission industry, has been increasing complexity; for example, the number of receipt points on the NOVA system has increased from 287 in 1975 to 771 in 1989. The number of major delivery points increased from 16 to 122 over the same period. It is clear that this trend involves an intensive program of meter station and lateral construction, with an attendant need for more productive means of welding smaller-diameter pipe and assemblies. A significant portion of NOVA's welding research and development effort involves these applications, and some specific examples will be given.

The last decade has seen an important move in the pipeline industry towards a fitness-for-purpose approach to the assessment of weld imperfections, culminating in the adoption of an alternative standard for primary inspection in the 1986 edition of the Canadian gas pipeline code, CAN/CSA-Z184-M86. This development has further encouraged the adoption of an engineering, rather than craft, approach to field welding, since the reproducibility of procedures from weld to weld becomes a key issue. In addition, the adoption of such standards has raised some significant concerns regarding the capabilities of traditional non-destructive inspection techniques. Some developments in this area which are beginning to see field application will be briefly described.

## 2. Hydrogen-Assisted Cold Cracking in Shielded Metal Arc Welds

Hydrogen-assisted cold cracking has been the most disruptive, if not the most statistically prevalent, quality problem associated with traditional, manual girth welding using cellulosic electrodes.

It has been understood for many years that the factors controlling cracking behaviour are hydrogen content, tensile stress and material susceptibility. However, the specific sources, timing and local magnitude of the stresses involved in pipeline welding were studied in detail only in the late seventies and early eighties (Higdon et al. 1980, North et al. 1982). As regards material susceptibility, there has been a constant evolution in pipeline steel formulation and production technology, in response to escalating demands on mechanical performance and increasing wall thickness. As a result, welding procedures which provided adequate safeguards against cracking in the seventies may no longer be sufficient today. Particularly as the use of higher-strength steels (Gr. 550 or X80) and/or heavier wall thickness becomes more prevalent, reliable means of predicting cracking susceptibility and developing procedures will remain as necessary as ever.

There has also been a change in the way in which the effect of chemical composition is assessed in the Canadian pipeline codes, so that many of the old empirical guidelines can no longer be applied. This has led to sustained research efforts to relate cracking behaviour to chemical composition. At NOVA, as well as in cooperatively-sponsored research, small-scale restraint tests have been used to relate critical preheat to prevent cracking, under standardized welding conditions, to chemical composition. Earlier full-scale testing, together with practical experience, allows field performance to be related to laboratory results (North et al. 1982); because preheat temperature is the only important field variable which can be changed (within limits) without affecting productivity or operating practices, it is a particularly useful yardstick.

Small-scale, WIC restraint cracking tests were first used by NOVA to assess the effects of some apparently minor changes in chemical composition which were not reflected in any carbon equivalent (Rothwell et al. 1986). This was followed by a more comprehensive series of tests, aimed at estimating the range of validity of the newly-adopted CSA carbon equivalent. This work indicated that the modified WIC restraint test, with a standard pipeline bevel, accurately reproduced the root-pass cracking site observed in the field when welding with high-strength (E55010 and above) electrodes; cracking initiates in the weld metal, at the toe region (Figure 1), and not in the internal, root region of the HAZ. In these circumstances, it is reasonable to seek a correlation of critical preheat temperature with weld metal chemical composition, rather than with that of the base metal, and Figure 2 indicates that a linear relationship was, in fact, observed. Of course, using electrodes with closely similar compositions, and fixed welding parameters, so that dilution and transfer of alloying elements are constant, a correlation with base metal composition would still be expected, and critical preheat, for a given consumable and welding procedure, can thus be predicted on that basis. A range of materials of industrial origin have been tested, and for most, rational explanations for any variation from the base-line correlation can be found in terms of specimen thickness, or of variations in chemical composition which are not accounted for in the carbon equivalent (Figure 3). For example, points representing thinner materials generally lie below the 12 mm correlation, as do those for steels with particularly low nitrogen contents or with titanium as the only microalloying element. A few materials have been found to have critical preheat temperatures significantly higher than predicted, and no acceptable explanation has yet been identified. This anomaly is of some concern, since related field problems have been encountered, and is the topic of on-going work.

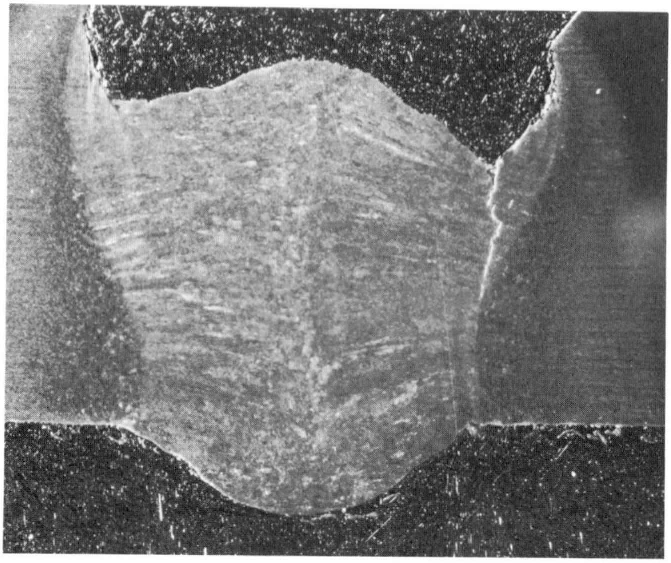

**Figure 1: Typical weld-metal cracking in high-strength root pass**

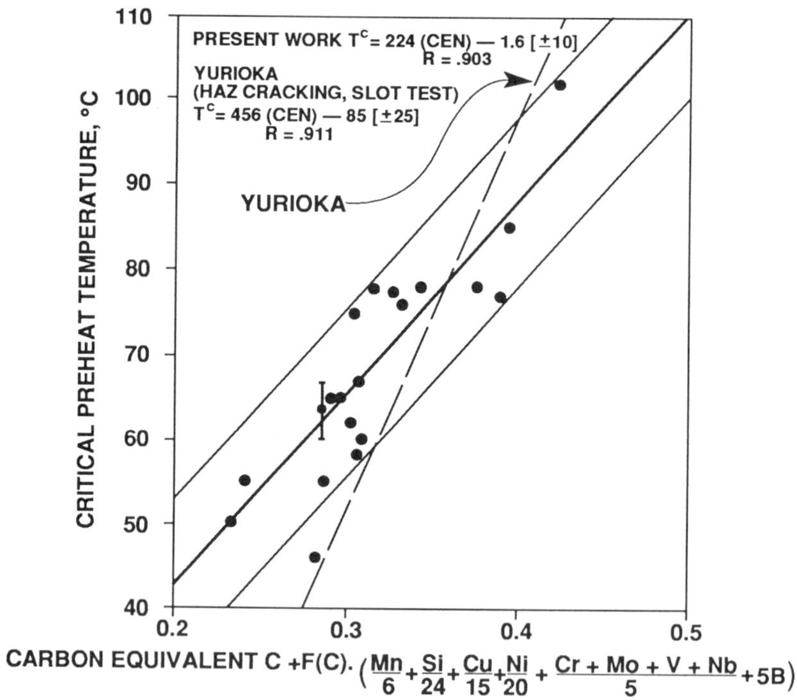


**Figure 2: Variation of critical preheat temperature with weld metal carbon equivalent**

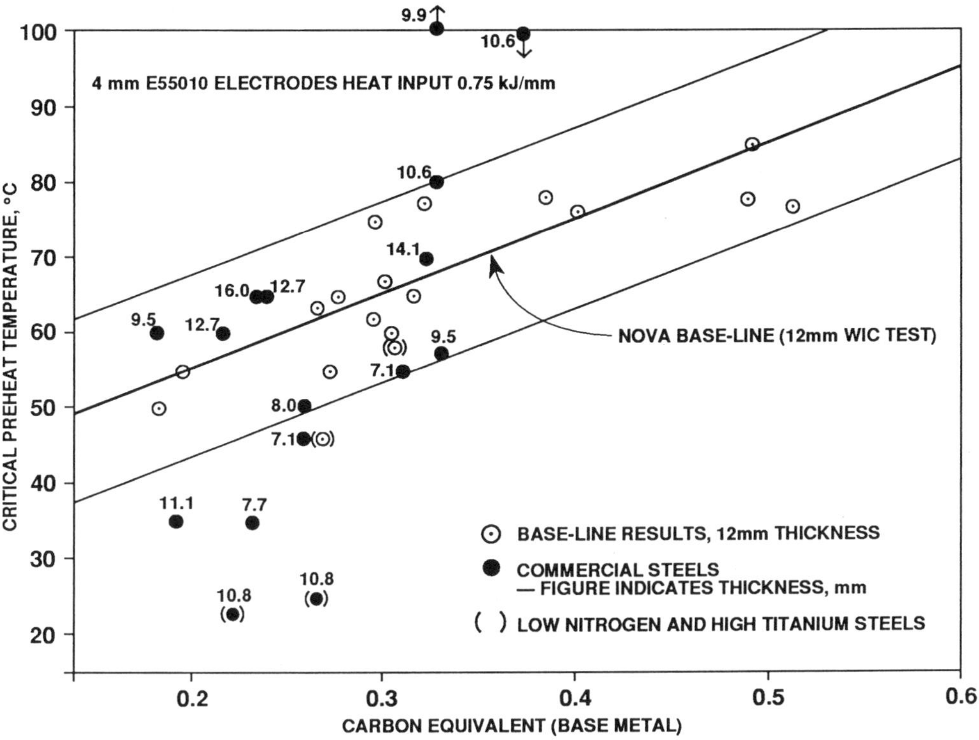


**Figure 3: Variation of critical preheat temperature with base metal carbon equivalent**

The preceding discussion has dealt with the situation in which the root pass cracks within a short time of deposition, as a result of the stresses arising from pipe movement, concentrated by the geometry of the incomplete weld. The effects of geometrical factors, including those of misalignment and the ratio between wall thickness and root pass thickness, have been investigated in some detail (Higdon et al. 1980). It is believed that this scenario accounts for the vast majority of all cracks in main-line welds, and that the regular practice of carrying out radiography within hours of weld completion will normally be successful in detecting such cracks. Certainly, the negligible number of hydrotest or service failures arising from girth weld defects is inconsistent with the presence of any significant population of undetected cracks. Recently, however, a situation has been encountered in which cracking has occurred in welds of a highly self-restrained nature (deep repairs and tie-in welds) after an apparent delay of several days. Such a phenomenon is highly disturbing, since it calls into question the premises on which construction and inspection schedules are currently based. The topic is the subject of active research in Canada at the present time.

## 3. Mechanized Gas Metal Arc Welding

### 3.1 Economic and Quality Factors in Conventional Mechanized GMAW

As was mentioned in the Introduction, main-line welding, being a highly repetitive process, should benefit greatly from mechanization. However, there is not a large choice of processes capable of producing over eighty fixed-position welds per day in pipe up to 1219 mm or more in diameter and having a wall thickness up to 19 mm. The CRC Automatic Welding mechanized GMAW process meets these requirements, and was first applied to a NOVA project in 1971; in total, over 1600 km of large-diameter pipelines have been constructed for NOVA using this or similar processes. Economic analysis of closely similar projects constructed in the same time frame suggests that the mechanized process can be at least twice as productive, in terms of man-hours per weld, as the conventional manual process; examples drawn from some major projects built in the early eighties are shown in Table 1. In addition, as a result of its intrinsically-low hydrogen content, GMAW is essentially immune to cold cracking. The typical process-related defects, porosity and lack of fusion, are much less disruptive, and can be dealt with rationally on a fitness-for-purpose basis, as will be discussed later. Their incidence can also be significantly reduced by process improvements.

Table 1. Productivity on Comparable SMAW and Mechanized GMAW Projects

| Project | Process | OD mm | WT mm | Manpower (welding) | Production (welds/day) | Productivity (man-days/ weld) |
|---|---|---|---|---|---|---|
| Northern Borders | Manual | 1067 | 15 | 217 | 72 | 3.01 |
| Foothills | Mech | 1067 | 12 | 164* | 171 | 0.96 |
| TCPL | Mech | 1067 | 10.3 | 192* | 160** | 1.20 |

* Two crews
** Winter construction

The mechanized welding approach currently used in Canada involves an internal root pass and special, field-machined bevels. As a result, equipment costs are relatively high, and economic viability is not achievable for short projects. There is a need for a simpler (though possibly less productive) mechanized process to deal with such projects, and research in this area is in its early stages.

The existing mechanized GMAW process uses the short-circuiting metal transfer mode to achieve all-position capability. As a result, the choice of shielding gases is restricted to those containing a high proportion of carbon dioxide. This imposes an intrinsic limitation on the toughness which can be achieved, since weld-metal oxygen contents range from 400 ppm upward. An early study of GMAW consumables found that some popular wire combinations gave marginal toughness, particularly in the context of a fitness-for-purpose or engineering critical assessment (ECA) approach (Dorling and Rothwell, 1984). While a judicious choice of wire led to acceptable toughness for the application, further improvements were felt to be necessary.

In addition to this limitation, the short-circuiting mode of metal transfer is prone to side-wall fusion defects, which can only be avoided by strict attention to arc positioning during fill-pass welding. Finally, in order to achieve adequate productivity in the hot pass, arc currents and travel speeds must often be at the limit of an acceptable operating envelope. If any material factors are present which compromise arc stability, unacceptable levels of spatter and poor bead profile will result.

All of these limitations can be addressed by the use of current pulsation to produce controlled spray transfer at mean currents which still allow all-position welding. The next section describes a systematic effort, which has been undertaken over several years, to optimize pulsed GMAW for pipeline and related applications.

### 3.2 Development of Pulsed GMAW for Pipeline Applications

GMAW in the spray regime offers a more controlled mode of metal transfer and may be operated with a wide range of shielding gases. However, the threshold current for spray transfer is too high to support all-position capability. It was recognized over twenty years ago that current pulsation provided a means of extending the range of spray transfer to lower mean currents and melting rates, and thus of moving beyond the flat and horizontal/vertical positions (Daggett and Zircher 1970). However, it was only in the eighties that advanced solid-state power sources were developed which were sufficiently flexible to allow a systematic optimization of pulse parameters (Allum 1983); equally important, from the pipeline industry's point of view, is the availability, at reasonable cost, of pulsed power sources robust enough to survive life on the right-of-way.

The principles by which arc stability and metal transfer in pulsed GMAW can be optimized were widely studied during the eighties in the UK, and some fundamental guidelines were developed (Amin 1983, Allum 1983, Quintino and Allum 1984, Smati 1986). Nevertheless, for each specific application, optimum welding conditions must be established, and these must be tied in to an appropriate synergic logic so that necessary changes in wire-feed speed/mean current for different regions of the weld do not disrupt the ideal metal transfer conditions. Because of the large number of variables in pulsed welding, a systematic approach is necessary; the relatively narrow operating window and impossibility of ad-hoc adjustments in arc length make mechanized, fixed-position girth welding in a steep-walled groove a particularly challenging application. Despite this, early work, using an available power regulator with less-than-ideal pulse characteristics, demonstrated the potential benefits to be achieved; for example, CTOD values at

-5°C were essentially doubled by the change from 100% $CO_2$ shielding gas to a 90% Ar-10% $CO_2$ mixture which the pulsed technique permitted (Ditschun et al. 1986).

The first systematic study carried out by NOVA of the effects of pulse structure on arc characteristics, metal transfer and weld metal properties was begun in 1985, and addressed uphill welding of relatively thick-section materials (Dorling et al. 1986). The principal potential application would be in large-diameter compressor station piping construction; earlier experience had indicated the severe limitations of conventional, manual techniques for wall thicknesses above 25 mm, both quality and productivity raising concerns. In fact, mechanized, gas-shielded flux-cored arc welding had been applied to station construction as early as 1981; because of the limited range of available consumables with true all-position capability, however, some further improvements in toughness were felt to be desirable (Huntley et al. 1984).

This initial investigation involved combinations of eight 1.2 mm diameter wires and three shielding gas mixtures of low oxygen potential. Earlier work had indicated that optimum metal transfer was achieved when one droplet, of a diameter similar to that of the wire, was detached close to the trailing edge of each high-current pulse (Figure 4), being transferred to the weld pool during the relatively quiescent, background period (Allum 1983).

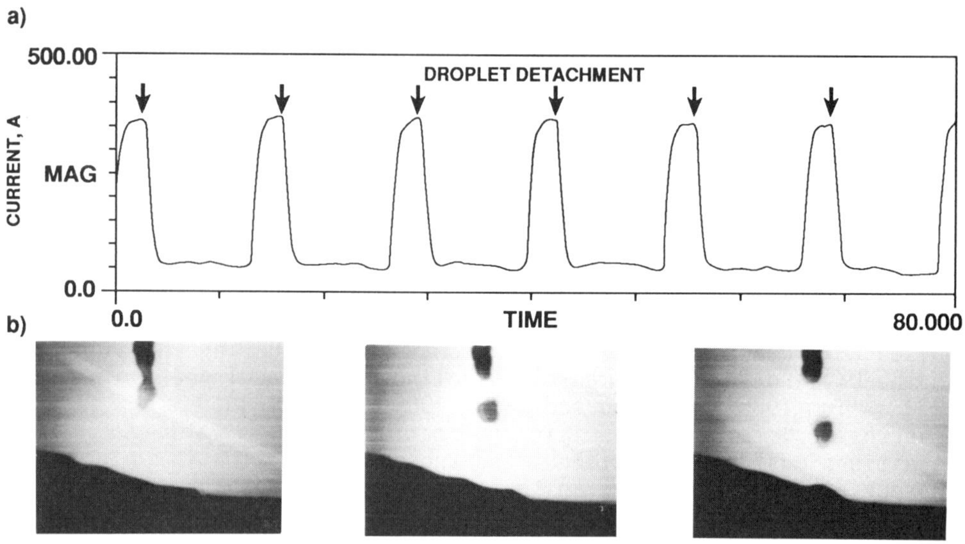


**Figure 4: Ideal metal transfer conditions at $I^P$ = 350 A and $t^P$ = 3 ms a) oscillographic trace of arc current, showing droplet detachment at trailing edge of peak; b) high-speed video**

The onset of these conditions was governed by a hyperbolic relationship between peak current and duration $(I_p)^n t_p$ = const, where the exponent n is in the range 2 - 2.5 (Allum 1983, Amin 1983, Smati 1986). NOVA work, using analysis of oscillographic traces and high-speed video recording, confirmed this general pattern, but indicated that the range of peak time before this ideal behaviour gave place to multiple droplet transfer was quite narrow, of the order of 1 ms at 350 A, and that the required pulse peak parameters could vary with wire and gas composition. It is significant that few commercially-available pulsed power sources have sufficient flexibility to allow ideal peak parameters to be achieved over a wide range of consumables and mean currents, and that the way in which many synergic systems respond to changes in mean current/wire-feed speed does not ensure that these conditions are maintained. In the NOVA work, optimum peak current and time for a specific application were determined experimentally and held constant thereafter. Mean current was as high as possible for the application, pool stability and bead shape being the limiting factors. Given the condition of one droplet per pulse, of a diameter equal to that of the wire, and the approximately linear dependence of burn-off rate on mean current, simple geometrical considerations dictate a constant ratio between mean current and frequency, for a fixed wire diameter. In effect

$$F \cdot 4\,\pi r^3/3 = W \cdot \pi r^2 \approx \text{const} \cdot I_m \pi r^2$$

where F is the pulse frequency (Hz)
r is the wire (and droplet) radius (mm)
W is the wire-feed speed (mm/s)
$I_m$ is the mean current (A).

The relationship between W and $I_m$ was investigated for a wide range of materials and electrical extension. For a contact-tip-to-work distance of 13 mm, which was felt to be a practical optimum, the constant in the above equation was found to be around 0.4. This gives a ratio between $I_m$ and F of 2, for a 1.2 mm diameter wire. The full pulse structure can thus be completed, as follows. Since the frequency F can be calculated from the mean current, and peak time is fixed, background time can be determined directly, ($t_b$ = (1/F) - $t_p$) and background current calculated from the relationship $I_m = F(I_p t_p + I_b t_b)$.

One additional refinement was found to be necessary when welding in a groove, in order to avoid process instabilities, and associated potential for fusion defects, as the oscillating arc impinged on the sidewalls. The best method of stabilization was found to be by varying pulse frequency in response to a voltage error signal. While the resulting variation in frequency/mean current ratio in principle leads to a change in droplet size, the variations caused by the relatively minor frequency changes involved (typically from 75 Hz to 90 Hz, for a nominal mean current of 150 A) would not be significant. This method of arc voltage compensation leads to a slight increase in mean current as the arc hits the sidewall, which is also beneficial from the point of view of fusion characteristics. It is significant to note that, once this technique had been adopted, no evidence of any lack of fusion at the sidewall or between passes was detected in any of the many butt welds examined (both radiographically and by sectioning) in this program.

To this stage, the research had concentrated on developing a full understanding of the influence of pulse structure on metal transfer and arc stability, and on a general investigation of burn-off relationships and fusion characteristics. Full procedures for 30 mm pipe and plate were now developed, and test welds were completed on plate using combinations of eight wires and three shielding gases. Most welds were made under 95 Ar - 5 $CO_2$; the anticipated low oxygen contents (200 ppm or less) were achieved, and Charpy energies of up to 100 J at -60°C were reached with the most successful consumables. These results were obtained despite the fact that no particular effort had been made to optimize heat input, which reached 5 kJ/mm for the fill passes. However, CTOD values at -45°C were generally marginal or poor, though some encouraging results were obtained by giving specific attention to root deposition techniques. In addition, the welds made under gases of even lower oxygen potential (88 Ar - 10 He - 2 $CO_2$ and 87.5 Ar - 10 He - 2 $CO_2$ - 0.5$H_2$), while showing the anticipated lower oxygen levels (150 ppm or less), gave no consistent benefit in toughness. It was clear that the optimum balance between the oxygen potential of the shielding gas and the intrinsic hardenability of the wire had to be sought on a systematic basis. A further phase of the program was thus launched in 1986, to study the effect of shielding gas composition on arc stability, metal transfer, fusion characteristics and mechanical properties (Thompson et al. 1988). On the strength of the experience already gained, it was considered appropriate, at this stage, to incorporate the more demanding (from a process point of view) high-speed, downhill, narrow-groove process appropriate for mainline girth welds.

The two electrodes chosen, on the basis of the previous work, were a relatively low-hardenability, C-Mn-Si-Ti wire of 0.9 mm diameter for the low heat input, downhill process, and a more hardenable, C-Mn-Ti-B wire of 1.2 mm diameter for the higher heat input, uphill process. The method of investigation adopted began with the optimization of oxygen potential for microstructure and toughness, using binary Ar-$CO_2$ mixtures. For the downhill process, toughness was found to show a relatively broad peak, in terms of both Charpy and CTOD values, at $CO_2$ contents from 12.5 to 17.5% (Figure 5). For the uphill process, a similar peak occurred between 7.5 and 12.5% (Figure 6).

The next stage of the investigation involved the determination of the effect of variations in gas composition on fusion characteristics, at the optimum oxygen potential. The influence of helium-argon ratio, and of small amounts of oxygen and hydrogen, was studied using bead-in-groove configurations which allowed fusion characteristics to be optimized using realistic geometries, for the two applications. At this stage, it was also necessary to re-determine acceptable pulse parameters for the downhill process.

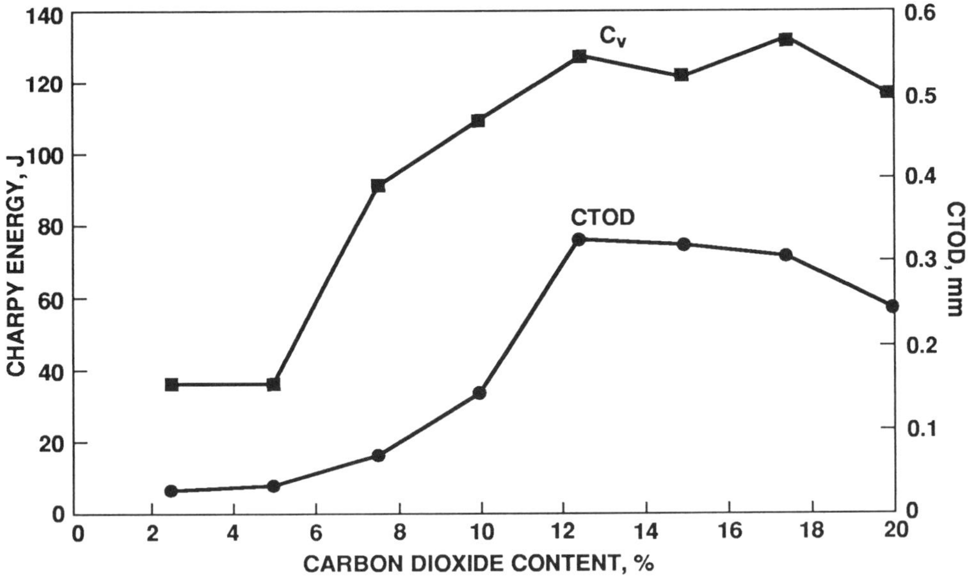


**Figure 5: Effect of increasing oxidation potential on Charpy energy and CTOD at -5°C in downhill welds**

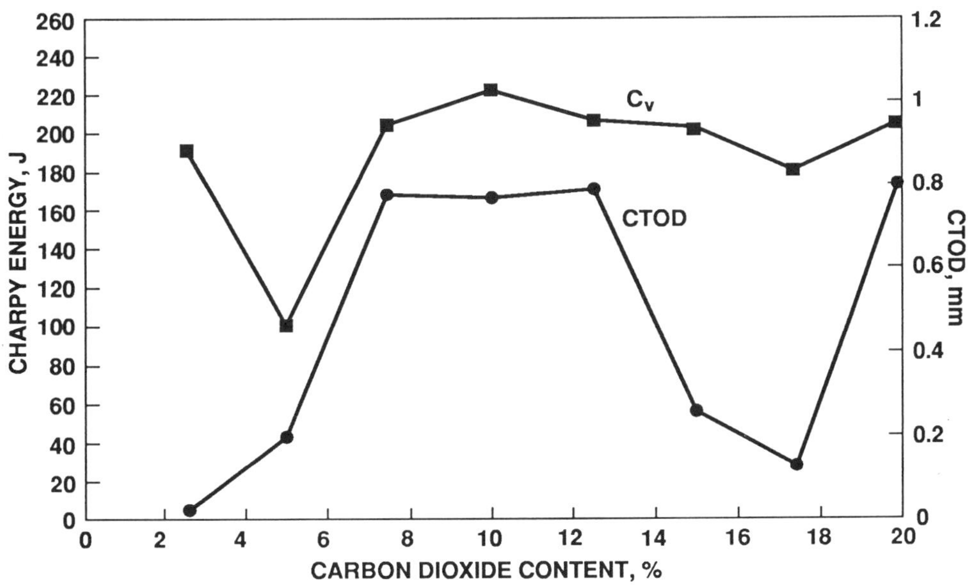


**Figure 6: Effect of increasing oxidation potential on Charpy energy and CTOD at -45°C in uphill welds**

Earlier work, on an empirical basis, had suggested that an $F/I_m$ ratio of 0.6 was appropriate for 0.9 mm wire (Quintino and Allum 1984). In pipeline welding, however, rather high mean currents are necessary in the hot pass (up to 230 A) in order to keep pace with the multi-head root, and geometrical considerations of the type outlined above quickly indicated that a somewhat larger droplet size must be tolerated, if ideal pulse peak characteristics are to be maintained. A droplet diameter of around 1.2 mm, with an $F/I_m$ ratio around 0.7, was found to give stable arcing conditions up to the necessary mean current levels.

The addition of helium significantly affected fusion characteristics. For the downhill process, 82.5 Ar - 12.5 $CO_2$ - 5He gave optimum performance, while 80 Ar - 10 $CO_2$ - 10 He was preferred for the uphill process. The addition of 0.5% hydrogen had no detectable effect on fusion characteristics; the substitution of carbon dioxide by 2% oxygen (82 Ar - 10 He - 6 $CO_2$ - 2$0_2$) significantly enhanced process stability, metal transfer and interpass fusion in uphill welding, while no obvious benefits were obtained by the addition of oxygen in downhill welding.

On the basis of these observations, the combinations of 82.5 Ar - 5 He - 12.5 $CO_2$ and 82 Ar - 10 He - 6 $CO_2$ - 2$0_2$ were selected for final procedure development, for the downhill and uphill processes, respectively. Operating envelopes for stable metal transfer, in terms of pulse parameters, were re-determined, for the specific wire/gas combinations, and qualification welds were made in 30 mm Lloyds LT60 plate (uphill) and NPS 42, 19 mm WT, CSA Z245 Gr. 483 pipe (downhill). The anticipated excellent arc and fusion characteristics were attained, together with very satisfactory Charpy and CTOD toughness (Table 2); the toughness obtained in the uphill weld suggested that some further fine-tuning of oxygen-carbon dioxide ratio may be worthwhile (compare Figure 5).

Although the deployment of pulsed GMAW in the field involves questions of economics and equipment availability, as well as weld quality, it is worth noting that the downhill process was used in 1989 for hot-pass deposition on a NOVA pipeline project where a potential for arc stability problems existed; performance and equipment reliability were entirely satisfactory.

Table 2. Charpy Impact Energy and CTOD Values for Procedure Qualification Welds

| Process | Test Temperature °C | Charpy Energy J | CTOD mm |
|---|---|---|---|
| Uphill | -45 | 193, 204, 239 | 0.28, 0.23, 0.32 |
| Downhill | -5 | 114, 120, 124 | 0.27*, 0.29*, 0.31* |

* Indicates $\delta_m$; all other values $\delta_u$

## 4. Process Developments for Small-Diameter Pipelines and Assembly Fabrication

### 4.1 Small-diameter Pipeline Construction

At present, all small-diameter mainline construction (below NPS 20) on the NOVA system is carried out using the manual SMAW process. The particular benefits offered by "one-shot" welding processes led, in 1986, to a program of work to assess the viability of magnetically-impelled arc butt (MIAB) welding for the construction of small-diameter, cross-country pipelines.

Developed for the joining of thin-wall steel tubing, MIAB welding is a high-temperature forge-welding process, which, in general, exhibits weld characteristics similar to other hot forge techniques such as flash, friction and high-frequency resistance welding. The essential difference is in the way in which heat is generated prior to the application of the forging force. With the MIAB process, the square-edged tube faces to be joined are separated by a small gap and a welding arc is established across the gap. A static, radial magnetic field is superimposed in the gap which causes the arc to move around the tube ends as a result of the interaction of the arc current and the magnetic field. The speed of the arc is very high (150 m/s or greater) and results in uniform heating of both tube faces. After sufficient heating time, the tubes are rapidly forged together to produce a solid phase bond with a characteristic flash or upset.

The major advantages claimed for MIAB welding over the alternative forge welding processes are:

- high welding speed (11 s welding time for NPS 4, 3.2 mm W.T.);
- low energy consumption;
- limited material loss;
- lower spatter;
- ease of automation.

These benefits result from the fact that the arc is an efficient heat source and that only axial movement of the parts to be joined is required during welding.

The process has been exploited mainly in the European automobile industry to weld a range of carbon and low-alloy steel components such as car and truck axles, drive shafts, shock absorbers and gas-filled struts. All of the commercially-available machines are stationary, floor- or bench-mounted units into which the parts to be joined are fed.

A feasibility study, conducted at The Welding Institute on NPS 4, 3.2 mm W.T., Grade 290 pipe, showed that welds could be made that consistently met the mechanical test requirements of the CSA Z184 standard over a wide range of welding parameters, demonstrating that the process had the potential for further development. Since that time, work has progressed on the design and construction of a fully-automated, portable MIAB welding machine with the capability of joining, in the field, pipe sizes from NPS 3 to NPS 12.

Girth welds produced by the MIAB welding process are of an entirely different character from conventional fusion welds; the bond line is perpendicular to the pipe surface and, as the weld is completed in "one shot", a single defect could, conceivably, penetrate the pipe wall. Weld defects typical of the MIAB welding process are nonmetallic inclusions in the form of thin films at the bond line; these films or "flat spots" are not easily detected by conventional radiography or ultrasonic examination and, in the high-volume manufacturing industries that utilize the MIAB welding process, weld quality is based almost entirely on process monitoring and control. The development of a process monitoring/adaptive process control alternative to NDE is part of the current research program. Nevertheless, some form of confirmatory NDE is required during the development of process procedures, and it is also anticipated that NDE will be required on initial production applications of the MIAB welding process to demonstrate the reliability of the process monitoring and control approach. This is also under investigation.

4.2 Small-diameter Assembly Fabrication

Until recently, small-diameter pipe assemblies were fabricated, in the shop, using GMAW for the root bead and low-hydrogen SMAW for the fill and cap passes, all in the 1G, or "roll-welding", position. With the considerable increase in pipeline activity, NOVA's fabrication shop became loaded to capacity; shop congestion precluded any significant addition of staff to meet projected increases in work load. The productivity of the welding staff was considered to be near optimum, for the manual welding and manipulation equipment installed at that time. Repair rates had been minimized and would have been expected to be adversely affected by an increase in hours of work. The proposed solution was to introduce new welding technology, and the combination of semi-automatic GMAW and FCAW is now well established.

Productivity evaluations led to the conclusion that small-diameter pipe assemblies can be welded two to four times faster with semi-automatic GMAW and FCAW processes than with manual SMAW procedures. The development and application exercise included equipment and consumable identification and selection, the development and qualification of fit-up, tacking and welding procedures, which would produce the required weld metal and heat-affected zone properties for the particular material and intended service of the components being welded, and welder training and qualification.

As the semi-automatic welding procedures were being developed and introduced, evaluation of a pipewelding system developed by the Industrial Materials Research Institute (a laboratory of the National Research Council of Canada) commenced. The system was specifically designed for the roll welding of small-diameter pipe using the GMAW process. Automatic completion of a full-penetration, unbacked, root bead is achieved by the use of a vision system; a video image of the arc and weld pool is processed by a computer and a fixed spatial relationship maintained between the tip of the GMAW wire and the contours of the weld pool, through adaptive control of current, voltage, travel speed and oscillation. Only the root bead is fully automated; fill and cap passes require manual tracking. Evaluation of a prototype system demonstrated that small-diameter assemblies could be completed three to four times faster with the IMRI system, compared with manual SMAW procedures. However, of the 176 production welds completed, 75% revealed porosity on radiography, with 38% requiring repair. The porosity occurred primarily at root bead start-stop locations. Since that time, an industrial welding system based on the IMRI technology has been designed and developed by Tecnar Automation of Montreal. With modified root bead welding control logic, based on NOVA experience with the mechanized GMAW pipeline welding process, the system can now consistently produce welds that meet the quality requirements of the CSA and ASME codes, and serious consideration is being given to the installation of the system in NOVA's Spruce Grove fabrication facility.

## 5. Structural Integrity and Detection of Defects

In addition to requirements for improved productivity, increased emphasis has been placed in recent years on the design aspects of pipelines, and on their ability to perform in a wide range of conditions. This range of operating requirements has led to improvements in the materials being used for line-pipe, in the consumables being used for field welding and in welding processes. The acceptance criteria for imperfections in the field girth weld, however, have been based on workmanship standards. These standards have been shown by experience to be safe and conservative for conventional welding processes, but they have no basis in design, material properties or toughness. In order to take full advantage of improvements in these areas, and of welding process changes, increased use of alternative acceptance criteria for imperfections, using engineering critical assessment (ECA), is being made. One key aspect in the application of an ECA is obviously the ability to detect the defect (and size it) using non-destructive testing.

Non-destructive testing is an important part of the pipeline construction process, providing the information used to control girth weld quality according to the appropriate standards. Cost-effective pipeline construction demands reliable and rapid inspection and currently depends on X-ray radiography. Radiography, however, is recognized to have some inherent limitations for the detection of inclined, planar defects (Lumb 1982). As was mentioned previously, current mechanized welding systems, using short-circuiting GMAW, are more prone to planar defects than is SMAW, and therefore require a different approach to non-destructive examination.

Inspection of mechanized girth welds to reliably detect planar defects can be accomplished using ultrasonic techniques. In order to keep pace with the welding production, the system has to be mechanized. Initial evaluation of such a system by NOVA in the early eighties showed that mechanized ultrasonic inspection was viable, but some problems arose because of both missed imperfections and false indications. In order to overcome these problems, NOVA, in collaboration with RTD, undertook a series of research programs, which culminated in the successful application of mechanized UT to a series of field projects for both NOVA and TCPL, in 1989, covering 350 km.

The problems of missed imperfections and false indications are considered to be related to the characteristics of the ultrasonic beam used, and to the design of the inspection geometry. In addressing these deficiencies, the approach was to use specialized beams to detect specific discontinuities. In order to achieve a large echo response the ultrasonic beam geometry was selected to be normal to the plane of the expected imperfections in the root region, and as close as practical to normal in the other regions. A typical configuration is shown in Figure 7 for 8.4 mm wall thickness. A similar array is used for both sides of the weld, and the total weld is evaluated in one circumferential scan. Accurate indexing to the original weld preparation is essential, to achieve maximum sensitivity in the root. Detailed calibration blocks specific to each weld project, wall thickness and configuration are required, as well as specific analytical techniques to eliminate root geometry effects. This attention to detail was critical to the successful application of the technology to field projects. The ability of the technique to provide reasonable assessment of defect depths is also important in the successful implementation of engineering critical assessment for onshore pipeline projects, with the added benefit of assured structural integrity.

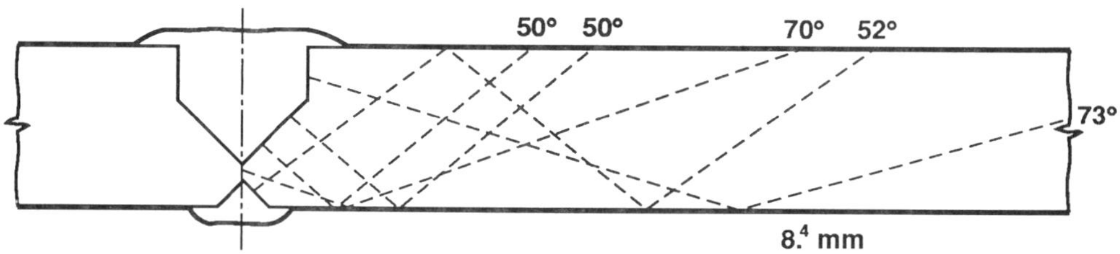


**Figure 7: Probe placement and beam geometry for mechanized UT of mechanized girth welds in 8.4 mm W.T. pipe**

## 6. Conclusions

In many respects the pipeline construction industry is very conservative, and welding techniques developed many years ago are still in use today. As new materials and design approaches are developed, and economic pressures increase, methods which revolve around craft skills and arbitrary workmanship standards become less and less appropriate. Any difficulties which arise are exacerbated by the strongly cyclic nature of the business, with its related effects on the stability of the work-force. There is still a need for "defensive" research, aimed at problems which arise with traditional welding methods, but more and more, given the intrinsically repetitive nature of cross-country pipeline welding, new, mechanized and automatic approaches will be sought, in which the quality of welds, and the way in which it is assessed, are related to prior engineering decisions, rather than to individual craft performance in the field. Some examples of research and development activities in both categories (defensive and innovative) have been given; while these have been drawn essentially from the authors' personal experience, and thus are far from comprehensive, it is hoped that they provide an authentic flavour of the directions in which the industry is moving, on a world-wide basis.

## References


Allum, C.J. (1983) MIG welding, time for re-assessment. Met. Cons., 15(6), 347-353.

Amin, M. (1983) Pulse current parameters for arc stability and controlled metal transfer. Met. Cons., 15(5), 272-278.

Daggett, E.H. and Zircher, W.E. (1970) New developments in pulsed spray welding. Weld. J., 49(10), 780-787.

Ditschun, A., Zajaczkowski, B. and Dorling, D.V. (1986) Pulsed FM-GMA Welding, in Welding for Challenging Environments, Pergamon Press, Toronto, pp 21-29.

Dorling, D.V. and Rothwell, A.B. (1984) The toughness of mechanized gas metal arc welds, in Welding in Energy-Related Projects, Pergamon Press, Toronto, pp 379-388.

Dorling, D.V., Huntley, R.M., Geerligs, J. and Rothwell, A.B. (1986) Pulsed gas metal arc welding of thick-section steels for low temperature application. NOVA Corporation of Alberta/CANMET unpublished report.

Higdon, H.I., Weickert, C.A., Pick, R.J. and Burns, D.J. (1980) Root pass stresses in pipeline girth welds due to lifting, in Pipeline and Energy Plant Piping: Design and Technology, Pergamon Press, Toronto, pp 263-272.

Huntley, R.M., Dorling, D.V. and Rothwell, A.B. (1984) Pipeline girth welding using the flux-cored arc welding process, in Welding in Energy-Related Projects, Pergamon Press, Toronto, pp 85-94.

Lumb, R.F. (1982) Non-destructive testing of pipeline girth welds in the 1980s. Br. J. NDT, 24(7), 199-207.

North, T.H., Rothwell, A.B., Glover, A.G. and Pick, R.J. (1982) Weldability of high strength line pipe steels. Weld. J., 61(8), 243s-257s.

Quintino, L. and Allum, C.J. (1984) Pulsed GMAW: Interactions between process parameters. Weld and Met. Fab., 52(2) and (3) 85-89 and 126-129.

Rothwell, A.B., Taylor, D.A. and Glover, A.G. (1986) The effect of nitrogen on the field weldability of high strength line pipe, in Inclusions and Residuals in Steel: Effects on Fabrication and Service Behaviour, CANMET, Ottawa, pp 445-467.

Smati, Z. (1986) Automatic pulsed MIG welding. Met. Cons., 18(1), 38R-44R.

Thompson, T.S., Rothwell, A.B. and Dorling, D.V. (1988) The influence of shielding gas composition on pulsed gas metal arc welding of Arctic and offshore structures and pipelines. NOVA Corporation of Alberta/CANMET unpublished report.

# RESEARCH DIRECTIONS IN WELDED HIGH STRENGTH STEEL STRUCTURES.

R.M. DENYS
Laboratorium Soete, Rijksuniversiteit Gent, Belgium.

Abstract
A multiplicity of factors determine the fail safe behaviour of steel weldments. This paper identifies a number of key points which need to be addressed when higher strength steel weldments are used in fracture critical parts. In particular, it is shown that weld joint efficiency hinges on the interaction between the relative difference between weld/base metal yield strength and toughness. It is concluded that a quantitative assessment of this interaction is required for the effective utilization of "new generation" higher strength steels. Also, it is explained that a re-examination of certain testing requirements and the development of supplementary ones are required to take full advantage of the higher strength steels.
Key words : High strength steels, Weld metal yield strength, Overmatching, Undermatching, CTOD, Wide plate testing.

## 1. Introduction

Today's steelmaking and plate manufacturing techniques allow fabrication of 'low-carbon low (lean) alloy' steels which are tough and readily welded by conventional techniques. Many proprietary lean alloy low carbon steels are available in a series of strength levels obtained by Normalizing, Quenching and Tempering (QT) or by Thermomechanically rolling followed by accelerated cooling (TMCP). Vast tonnages of Normalised or TMCP structural steels having a minimum yield strength of 350 MPa (high strength steel) are being used in virtually every high duty construction. Steels ranging in yield strength from 420 to 500 MPa (higher strength steels) are receiving increased interest. The yield strength in the latter steels is obtained either by a QT or a TMCP heat treatment.

In selecting steel for high duty constructions two aspects are of special importance. These are weldability and fracture behaviour. Modern QT and TMCP steels can be readily welded. In particular, the low levels of adverse hardenability elements in modern steels have the effect of reduced HAZ crack sensitivity. Since TMCP steels permit a leaner chemistry than required for QT steels of equivalent yield strength, TMCP steels are usually considered as advantageous in terms of weldability. But, it should also be realized that the strength properties of TMCP steels depend on a complex and a critical combination of composition, rate of thickness reduction during rolling, rolling finishing temperature (thermomechanical rolling) and (on-line) accelerated (AC) or intensive (AcC) water cooling conditions following rolling.

Though higher strength QT and TMCP steels are very attractive, they may turn out to be recalcitrant in fracture behaviour because of their particular characteristics which are normally not considered to be significant. For example, as the low to very low carbon content of those steels has the advantage of making them easier to weld and thougher at low temperature, there is also the disadvantage that their tensile strength is reduced so that the yield to tensile ratio increases to levels for which little practical experience exists. Also, because of the complex rolling and cooling schedules, the TMCP manufacturing technique may produce plates which are considerably stronger than the minimum specified yield strength.

These and other problems to be discussed later demonstrate that, for instance, an adjustment of the weld consumable selection criteria may be required when the weld deposit properties have to match the plate properties. Therefore, the aims of this paper are to point out the basic characteristics of modern higher strength steels and to explain why these differences call for adjusted fracture control processes. It is further the intention to provide guidance on the pertinent material testing aspects in order to generate realistic, safe and meaningful steel and weld metal specification requirements.

## 2. Steel properties

Designers and structural engineers, unlike steelmakers, are mainly interested in how a steel performs under load. Often they accept blindly the steel available without giving due consideration to their microscopic and macroscopic qualities, i.e. many of the desired steel properties are so obvious to structural engineers that they accept that the specified properties comply closely with the minimum requirements. It is true that designers who rely on the current opinion that the specified properties are supplied accordingly are in error.

Modern low-carbon low-alloy steels differ from conventional "medium" carbon steels in many aspects. The differences, however, are not always appreciated, or in fact, unknown. So, it may be profitable to review the essential qualities of a modern low-carbon low-alloy steel.

Though it is impossible to consider the individual qualities of each type of high(er) strength steel in detail, it should be realized that modern steel plate manufacturing techniques cover a wide range of methods to obtain the desired yield strength properties, Fig. 1.

### 2.1. Yield strength properties.

The yield or proof stress is most important single parameter characterizing a material from the point of view of initial yielding. The yield stress alone is not sufficient to characterize deformation behaviour in the occurrence of defects. It is considered that the full stress-strain curve should be evaluated because post yield behaviour of steel, even under conditions of adequate ductility, is related to the yield to tensile strength ratio or the strain hardening capacity.

In perspective, it may be useful at this point to discuss the yield and tensile properties of today's steels. It can be verified that modern higher strength (i.e. QT or TMCP) steels have a higher yield to tensile ratio (YS/TS) than conventional -high strength- steels. The reason is that strenghtening in modern higher strength steels is obtained by special rolling conditions at low carbon levels. In particular, reduction in carbon content causes a marked decrease in tensile strength whilst the rolling conditions are

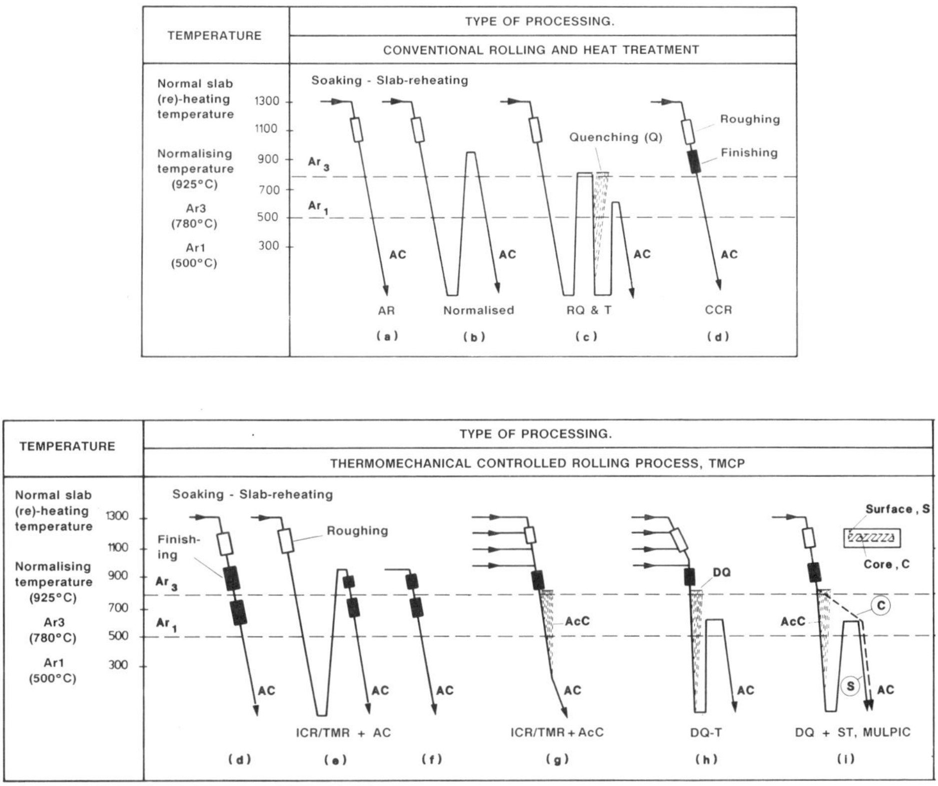


Fig. 1. Schematic profiles of conventional and thermomechanical steel plate rolling techniques.

adjusted to obtain the required yield strength with the end effect that the yield to tensile ratio rises sharply. This increase can produce serious difficulties because steels with such high YS/TS ratio have inadequate strain hardening capacity to redistribute local deformation. Where higher strength steels are used, it can therefore be propounded that the favorable experience gained in the application of (high strength) Normalised or TMCP steels can be badly distorted.

Previous observations are of more than academic interest. For example, it is hardly known how serious increasing YS/TS ratios affect fracture behaviour of (cracked) welded structures. On the other hand, high YS/TS ratio steels can most advantageously be utilized where the design stress is based on yield strength. If the design stress is the lowest of e.g. TS/2.7 or 0,2%.YS/1,6 (design criterion for thick walled pressure vessel, ISO - R831), the optimum YS/TS ratio is 0,60 implying that high YS/TS ratio steels cannot be used to the greatest advantage for this application. But, in terms of design-stress-safety-factor, it can be concluded that a higher safety factor will be reached for higher YS/TS ratio steels.

Having taken the decision to use higher strength steels, there is also the need to give due consideration to the actual yield strength properties. Most designers are generally ignorant about the fact that some types of TMCP steels have yield properties appreciably higher than specified.

A general evaluation of yield strength variability is a difficult subject since there is a large variety in the expedients adopted by the different steelmakers to obtain the specified strength, ranging from variations in alloy design to 'special' TMCP plate processing, Fig. 1. Thin plates, for example, show usually a rather wide variability of yield strength properties. In fact, it is quite possible that the actual properties may qualify for a higher steel grade than indicated by the guaranteed minimum properties. In this context, the user must also be aware that TMCP is a generic term that refers to a general class of proprietary processes, procedures and chemical compositions, i.e. all TMCP steels of the same strength are not created equal.

Since the designer uses the base metal's specified minimum yield strength as one of the criteria in weld metal selection it is imperative to appreciate the base metal's yield strength distribution. Thus, designers confronted with modern steels should know that weldments made with steels of the higher end of the yield strength distribution may contain undermatching weld metal.

Previous observations illustrate that it is of great practical interest to establish the distribution of yield strength properties. In particular, specifying a maximum yield strength is significant because it will help to select the correct weld consumables and welding parameters (see further). However, specifying narrow limits for yield strength such as SMYS + 80 upto 100 MPa maximum does not always find the ready acceptance of the steelmaker.

## 2.2. Toughness

Considerations of fracture control of steels to be used in welded structures are based on the prevention of fast fracture. Discrimination between steels as to toughness gave rise to numerous tests seeking to reproduce service performance in small scale laboratory tests. Research in several laboratories led to the development of the transition temperature concept in selection of steels in the medium to high strength ranges. This approach is still in vogue and the preferred tests used to determine transition temperature are (a) the Charpy V notch impact and (b) the Pellini NDT test.

The first Charpy V test requirements were obtained in the 1950's by correlating the characteristics of service failures and the Charpy V notch properties of steels involved in such failure. In the 60's, laboratory tests such as the notched and welded wide plate tension test were used to derive Charpy V impact requirements of C and C-Mn structural steels developed in the 1960 (Woodley & al. 1964). A re-evaluation of the conditions on which these correlations are based (Pisarski, 1986), additional research and wide plate test studies indicate that higher Charpy V impact properties may be needed for a correct rating of low carbon low alloy steels. In this connection, it should be mentioned that there is no unanimous agreement on the minimum test requirements.

In the absence of a coherent service performance or wide plate test data, most Charpy V notch impact test requirements for low carbon micro alloy steels are largely based on engineering judgment i.e. the impact energies are increased in proportion to the yield strength. In conditions where notch toughness is important, experience shows that this approach provides adequate requirements for 350 MPa yield strength steels. However, for higher strength ($\sigma Y$ > 450 MPa) steels some reluctance is observed to implement the same principle. The information arising from experimental studies, and particularly that from wide plate tests supports the opinion that the currently used Charpy V notch test requirements are not reliable, and that some other criterion than a proportionate increase in impact energy should be used.

If the existing proprietary structural steels are examined in terms of Charpy V impact properties, it may be concluded that low carbon low alloy steels have prominent toughness properties. Low carbon fine grained steels perform, in terms of Charpy energy, much better than conventional-old-type medium carbon level steels. Fig. 2 illustrates the characteristic shift of both the transition temperature curve and the increase of upper-shelf energy impact levels.

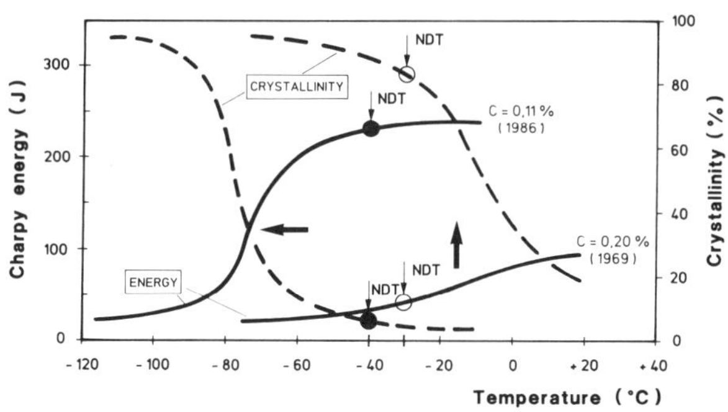


Fig. 2. Typical Charpy V impact energy and fracture appearance temperature transition curves and, related drop weight (NDT) temperatures for a conventional (C = 0,20%) and a modern (0,11 %) steel.

The comparison between the position of the Pellini Nil Ductility Temperatures (NDT) and the Charpy transition temperature curves illustrate some key points. In conventional steels, the NDT demarcates the limit between the lower shelf portion and the actual transition range. However, with modern low carbon steels there is a trend that the NDT correlates with the lower limit of the upper shelf portion of the Charpy transition curve, whilst the percentage fibrosity in the Charpy specimen at NDT temperature is much higher. This property could possibly be explained by specimen size effects; the Charpy specimen provokes a much larger temperature shift than the Drop Weight specimen. This finding suggests that the Charpy test specimen is not necessarily the best means for a representative rating of the transition temperature of low carbon steels.

It must be evident from the above described features of modern steels that there is scope for considerable research to establish Charpy V notch impact requirements for modern low carbon steels. In particular, the currently specified Charpy V impact requirements of steels with a yield strength of 450 MPa and beyond require a back up with full thickness wide plate tests.

## 3. Weldments properties

Obvious is the fact that in any welded structure the welds themselves are preferential sites for fracture initiation (note that the following considerations are focused on initiation behaviour). There are two regions i.e. the weld deposit and the HAZ where defects are likely to occur. The question is thus posed : to what extent can such defects be tolerated from the point of view of fail safe behaviour.

In modern higher strength steel weldments the situation is complicated by the fact that, because of the higher strength, fracture control is becoming a difficult subject where brittle fracture is a credible failure mode since the properties demanded are often on the fringe of currently available knowledge and technology.

Notwithstanding the favorable experience built up in the last decade, it is important to emphasize that the concept of fail safe design, when applied to higher strength weldments, embraces a number of novel aspects. The primary but well known concern is fracture toughness. In particular, when NDE techniques reveal flaws unacceptable by the particular code of construction it is often experienced, when additional testing is required, that the ultimate deformation behaviour does not reflect the excellent plate characteristics. However, it is an over-simplification to believe that the mechanical behaviour depends only on toughness (Denys, 1990). An additional consideration for higher strength steel weldments is certain to be the relative difference between weld and base metal yield strength. The reason for this is that the protection afforded to weld metal and HAZ defects in high strength steel weldments by the overmatching yield strength of the weld metal may be no longer available in higher strength steels weldments.

### 3.1 Toughness testing considerations

Of the many and various toughness tests for weldment evaluation, the notch (Charpy V notch impact) and the fracture (CTOD) toughness tests are customarily used to assess the toughness properties of the various zones in a weld (Pisarski & Walker, 1986). In what follows, the considerations will only be focused on CTOD since the recent emphasis on CTOD (HAZ and weld metal) toughness requirements.

### 3.1.1 CTOD toughness requirements

The CTOD approach is a popular method to evaluate weld joint integrity (Harrison, 1980). For the situation of elastic-plastic material behaviour, the CTOD concept requires a material toughness parameter, CTOD, and the conditions for fracture are derived from a relationship between CTOD, applied stress and tolerable defect size. That means that the assessment of weldments by use of fracture mechanics CTOD tests is an alternative to the Charpy V notch impact transition temperature approach outlined above.

Today, fracture toughness testing is centered on CTOD three point bend testing of a fatigue pre-cracked B x 2B (or B x B) specimens. However, this does not mean that the CTOD test is suitable as a production quality control test; the primary aim of the test is to measure the CTOD for use in defect assessments. Also, the results of CTOD tests can be used for ranking the response of a steel to the weld thermal cycle and welding procedures.

CTOD toughness requirements can be derived from the design curve when reasonable estimates of the input parameters are available. For critical applications, however, a specific assessment should be considered. For the situation of higher strength steel weldments, the difficulty is that existing CTOD toughness requirements are often duplicated without having a representative correlation established between the driving force for fracture and service or wide plate test data typical for today's modern steels and welding procedures. In most instances, there exists also a trend for over-specifying CTOD toughness requirements and this vogue is a natural outgrowth of both the statistical uncertainties associated with the use of the CTOD method and the insufficient knowledge of the fundamental deformation patterns occurring in higher strength steel weldments.

One of the reasons for these problems stem from the fact that the design curve was established by application of both simple fracture mechanics principles and correlation of the results of CTOD and wide plate tests produced in the 1970's (Kamath, 1981). The major problem with this approach is that no consideration was or could be given to the effect of the relative difference between weld and base metal yield properties i.e. the CTOD approach is essentially a single parameter solution, whilst the problem of higher strength weldments requires at least a two material property parameter approach.

### 3.1.2 Defect assessment

From a defect assessment viewpoint, the CTOD based fracture analysis of high constraint structures is rather well developed. However, CTOD fracture assessments for conditions of low constraint, low or high toughness and weld metal defects present considerable difficulty and provide often over-conservative tolerable defect size predictions. The problem is that the measured CTOD fracture toughness is sensitive to the geometry of the CTOD test specimen used, whilst it is not always possible to model the constraint conditions prevailing in the cracked structure.

At present, little information is available for establishing representative and safe CTOD toughness (and Charpy V notch impact) requirements for weld metal and heat affected zones. In higher strength steel weldments, both toughness and the relative difference between weld and base metal properties need to be incorporated.

When wider use is made of higher strength steel weldments, there will be also a greater need to coordinate toughness, defect size and service stresses. For example, unusual loading condition may lead to situations where the base and thus the weld metal is forced to bear plastic deformation. In order to ensure serviceability for such loading conditions, the conventional fracture mechanic concepts must be abandoned and alternative assessment procedures are needed.

—For all these and other types of problems, comparatively little research has been carried out and it should be noted that there is a paucity of full scale test data for solving these outstanding problems in a quantitative way.

### 3.2 Current use of wide plate test data

Although there is currently an increasing interest in the use of CTOD fracture mechanics based concept for the assessment of weld defects, it is also true there is an increasing interest to validate the CTOD based tolerable defect predictions with results of full scale or wide plate tests which are specifically designed to examine a particular defect or crack geometry. This approach recognizes that there is still interest in wide plate testing and that wide plate test data can be used to provide a direct and quantifiable measure of the fracture initiation characteristics of defective weldments (Denys, 1990a)

Today, however, there is some discussion amongst experts on the ultimate application of wide plate test results. Some experts prefer to employ wide plate test results in combination with CTOD fracture mechanics test results to quantify the degree of safety implied in CTOD design curve predictions (Pisarski & Walker, 1986). In other words, in this view wide plate test results are to be used to substantiate conclusions drawn from fracture mechanics tests and analysis. Others, consider wide plate test results as a suitable means to assess the structural implications of low fracture toughness properties as measured in Charpy V impact or CTOD testing (Denys, 1990). In this context, there is also some difference in opinion on the required failure stress/strain acceptance level as measured in a welded pre-cracked wide plate test specimen.

From an engineering point of view, these two types of assessment are quite different, however, both assessments merit consideration and are complementary. Actually, the preference for one or the other concept depends on the specified performance requirements, the objectives of the wide plate test and the response of the particular material to the test.

### 3.3 Mechanical properties of weldments

A fracture mechanics assessment of HAZ or weld metal defects is based on the assumption that the weld deposit is "stronger" than the base metal. The word stronger is used here to underline that the mechanical property requirements stipulated in most Codes and Standards are based on tensile strength and not on yield strength. It must also be recalled that the stipulated requirements are based on the specified minimum tensile strength.

—There is also another problem. Too frequently, designers expect that the results of a transverse tension test represent the properties of all production welds. Whilst it is not the intention of this paper to describe in detail the deficiencies of the transverse tension test, some comments on the currently used Codes and Standards which deal with the strength of the weld deposit would be useful. For example, Codes and Standards do not recognize that :

(a) the yield strength is one of the most critical parameters in a fracture assessment

(b) the yield strength level derived from electrode classification testing may not be obtained in the actual weld joint.

(c) the actual mechanical properties of the base metal are greater than the minimum specified value (effect of statistical distribution).

(d) the yield strength of the weld deposit can be lower than the actual base metal yield strength.

The overall effect of the last observation is that, at a given applied stress level, an "undermatching" weld deposit will be working at a higher value of the ratio applied/yield stress than the base metal so that the applied strain will be concentrated in the weld deposit, Fig 3.

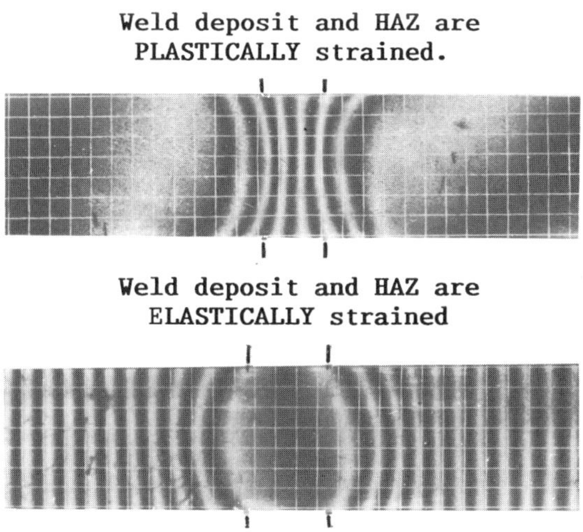


Fig. 3. Moire pictures illustrating the effect of weld metal yield strength undermatching (top) and overmatching (bottom). Applied strain : 1,2 %.

Further, when weld defects occur it must be understood that a higher toughness level is necessary to tolerate a defect of a given size than in the case for overmatching weld metal. An alternative approach which avoids severe toughness requirements is to select a compatible electrode that develops weld metal yield strength overmatching. Therefore, it seems logical that a minimum weld metal yield strength should be stipulated to prevent the situation where a weak weld (undermatching) would have to take all the deformation in the event of overloading.

As both yield strength and toughness of the weld deposit dominate the integrity a defective weldment, it is also clear that the treatment of weld defects will change according to the relative difference between weld and base metal yield strength. Thus weld metal toughness requirements and tolerable defect size predictions should be matched to the actual situation i.e. after considering the actual mechanical properties of weld and base metal or reasonable estimates thereof.

Defects in undermatching weld metal require a fracture assessment based on fracture mechanics and plastic collapse concepts. For defects in overmatching weld metal, both fracture mechanics and plastic collapse acceptance criteria can be applied, but there is also an incentive to base the assessment on the Gross Section Yielding requirement (Soete & Denys, 1976).

### 3.4 Weld metal yield strength evaluation.

One of the most trying problems in predicting weld joint integrity of higher strength steel weldments is the blind trust designers place in the properties of the supplied consumables. There are various grounds to admit that the pre-set (minimum) properties are not always met. Also the scatter, represented by the width of distribution, of the weld metal yield strength properties is higher in proportion than is customary in base metal yield strength properties. Many of the reasons for this scatter are well understood by welding engineers, and these have to do with small variations in wire, coating or flux composition and welding procedures. Note also that the engineering consequences of weld and base metal yield strength variability are not recognized in Codes and Standards.

In the absence of relevant information, however, it is worthwhile to give some consideration on the effects of weld metal (as well as of the adjoining base metal) yield strength variability so that designers know what to specify or what kind of testing requirement they have to implement to avoid the undesirable effects of weld metal undermatching.

The essence of the problem may be represented by a schematic diagram, Figure 4. Let A represent the distribution of base metal yield strength, while B is the distribution of weld metal yield strength, then the extent of overlap (interaction) between the two distributions is allows determining whether weld metal yield strength undermatching is significant. As shown, there is statistically a greater chance of having weld metal undermatching in higher strength steel weldments.

Because of specimen size effects it is clear that, in terms of the mechanical strength, the yield properties of an inhomogeneous weldment must be characterized, not by a comparison of the results of a single all-weld and a standardized (cylindrical) base metal test but rather by a representative composite (transverse weld) test specimen.

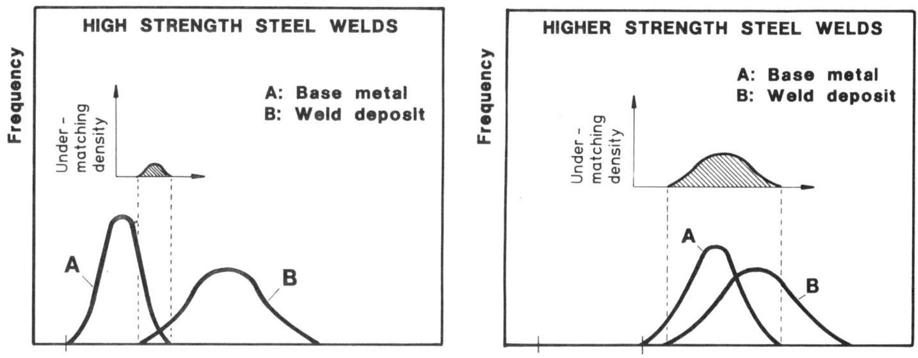


Fig. 4. Distributions of weld and base metal yield strengths.

In fact, the required information can be obtained by adjusting the testing procedure of the full plate thickness transverse tensile test.

By recording the load-elonation curve and assessing the (offset) yield strength of the composite specimen against the yield strength of the base metal derived from the load-elongation curve of a base metal test on the same specimen geometry, it is quite easy to verify whether or not the weld deposit yield strength is undermatching.

When the weld metal is undermatching, yielding will start in the weld deposit and the deflection of the load-elongation curve of the composite represents weld metal yielding. For matching and overmatching weld deposits, the yield point of the composite represents that of the base metal. However, for the situation of weld metal yield strength overmatching it is evident that the actual weld metal yield strength is not observable on a load-elongation record of a transverse tensile test. In such circumstances, an alternative measuring technique is necessary to determine the magnitude of the weld metal yield strength and thus the actual degree of weld metal overmatching (Denys, 1990).

A difficulty with this approach is that one single test cannot determine with certainty that all welds will meet the desired level of weld metal (over)matching. The primary problem which arises is that the distribution of the weld metal properties is not known. This lack of knowledge can be circumvented by the following approach. Instead of using test materials drawn at random from a range of possible yield strengths, it may be rational to define a characteristic of the sample i.e. the test material, and in particular the base metal, should be selected from the higher end of the yield strength distribution. For the weld metal, the problem is more difficult but, until more information becomes available, the approach should consist of a sensitivity study of the pertinent welding parameters i.e. it is the responsibility of the welding engineer to define and control the welding parameters to obtain reproducible weld metal properties.

## 4. Development of testing requirements for higher strength weldments

It is fitting to mention that the time is 'right' for a broad look at the subject of weld joint performance. Paramount in these efforts is the need to develop fracture performance requirements derived from suitable laboratory or structural tests.

Today, it is must be admitted that we do not always understand in a quantitative fashion the role of the relative difference between weld and base metal yield strength on the crack driving force for fracture. This comment can be further be developed taking into account the many innovations that have taken place in base and weld metals and in the type of service that are required of welded structures.

The relevant approach to tackle the problems consists of conducting large scale or wide plate tests. Although it is not very practical to conduct wide plate tests on a route basis, wide plate testing becomes important with increasing use of higher strength steels because wide plate test results are an essential element for identifying the interaction between the critical parameters dominating fracture behaviour.

For higher strength steel weldments, the approach should consist of a detailed examination of the deformation behaviour of a transversely loaded weld with weld metal defect so that the results of these studies could then be used for (re-)establishing production control specifications based on the results of small scale tests. The strategy to be followed would consist of :

(a) the development of an analytic model which relates the applied stress to the crack tip driving force for a crack of known geometry using the stress-strain characteristics of the composing parts of the weld,

(b) the validation of the predictions with actual test results obtained from a test configuration which enables a proper assessment of the interaction between weld and base metal deformation behaviour.

The next factor to be considered is related to the test performance requirement. When certain conditions with respect to instrumentation are satisfied (Denys, 1990a), the overall strain measurement can be used to establish a defect acceptance criterion. A consistent but simple and practical engineering means for assessing both elastic-plastic and plastic material behaviour is the Gross Section Yielding (GSY) concept. For defective welds, GSY is achieved when the gross section fracture stress exceeds the yield strength of the base metal.

The concept rests on the idea that when the material at the crack tip can strain harden enough to compensate for the missing cross sectional area in the plane of the crack, the applied plastic strain can be (uniformly) distributed all along the specimen length prior to specimen failure (Note that the GSY requirement is generalization of the 0,5 % (4 x yield strain) requirement established by Wells (1964) and which was developed to provide an answer for a particular problem related to local strain embrittlement in "old type pressure vessel steels").

When GSY is achieved, it is important to know that there are two possible paths to assess the wide plate test result :

(a) The test result can be used in defining directly fitness-for-purpose when a pre-set overall strain level is achieved. Such an approach, if well documented, does not require pre-qualification during fabrication when worst case situations are adequately modeled since the measured strains are often persuasive so that no further analysis is required. In other words, after-the-fact fitness for purpose evaluations are not needed when a proper engineering of representative test variables, before the start of production, is implemented.

(b) Provided that the weld metal overmatches the base metal in yield strength, it is possible to merge the well-established linear elastic and elastic-plastic fracture mechanics relationships with the requirement of GSY (Denys 1989). With undermatching weld metal, however, plastic strains will be concentrated in the weld metal. In this case, the Gross Section Yielding requirement can not be satisfied and reliance has to be placed on fracture mechanics and net section yielding or plastic collapse concepts for assessing 'undermatching weldments'.

## 5. Conclusions

1. It is a fact that the typical properties and problems of higher strength steel weldments are not always recognized by designers for the simple reason that designers rely on specifications which do not cover the intrinsic changes. If the full potentialities of low carbon higher Q&T and TMCP strength steels are to be realized, the development of a realistic basis for design against fracture is an urgent necessity.

2. The solution to the many known and unknown problems typical for higher strength steel weldments requires more test and development work so that more experience can be gained before starting work on novel constructions of novel steels. This approach necessitates also the development of test methods which incorporate the interrelation of the many parameters governing weld joint performance.

3. Designers must be aware of the fact that toughness testing is only one aspect of fracture control in higher strength steel weldments. It is a fact that we know little about the deformation aspects of defective welds in higher strength steels. If we are to apply higher strength steels which have to withstand difficult operation conditions, we have to consider the role of the relative difference between weld and base metal yield properties on weld joint performance. In particular, the future of higher strength steel weldments is bound up with the mechanical properties of the weld metal, and time is ripe to provide the means for predicting the interaction between toughness and yield strength properties of the composing parts in higher strength steel weldments. Thus, in addition to toughness testing requirements, there is a need for a review of Codes and Standards dealing with the evaluation of the relative difference between weld and base metal yield strength.

4. For the time being, information on the variability of both base and weld metal yield strengths would already be of great assistance to designers. This information, however, it not always available but the steel industry should be the source of statistical data on base metal yield strength properties, whilst it should also be their responsibility to narrow the range of variation through quality control. The tolerance to weld metal yield strength variations should be established as part of the welding qualification procedure. This process should define the acceptable ranges of heat input (welding parameters) necessary to meet the required level of weld deposit yield strength overmatching or other property requirements, such as toughness.

5. As a next step, the supply range of the yield strength properties of both weld and base metal should be standardized. This standardization should involve the specification of both a maximum base metal yield strength level and a minimum weld metal yield strength level to prevent strain concentrations (undermatching) in the weld metal region. Note that for less critical applications, less severe qualification requirements could be imposed. In this connection, note also that welds containing undermatching weld metal should carry the penalty of higher toughness requirements.

6. Turning again to the question of toughness requirements for higher strength steel weldments, it is evident that this subject needs also special attention. Two areas of uncertainty still exist. In the first place, the existing empirical correlations used to establish toughness requirement only work reliably within the range of conditions for which they have been developed. In the second place, there is also an urgent need to develop test requirements for lower temperature service. Unfortunately, as with most empirical correlations, interpolation of existing correlations is usually safe, but extrapolation is often hazardous. That is, it is technically not justified that toughness requirements established in the early 1960's are extrapolated without verification of these extrapolations in a representative situation. Therefore, in the present state of knowledge, the problem of toughness requirements for higher strength steel weldments needs more wide plate test results. From these results, low cost quality assurance requirements should then be derived.

## 6. Acknowledgment

The author wishes to thank A. Vinckier, director of the Lab. Soete, Gent University - Belgium, for permission to publish this paper. The financial support of IWONL and NFWO is also acknowledged.

## 7. References

Denys, R.M. (1989), Difference between small and large scale testing of weldments, An evaluation of the weld metal overmatching effects in relation to weld joint performance., The Welding Journal, 2,

1989 Research Supplement, p 33s-43s.
Denys, R.M. (1990), The Effect of HAZ Softening on the Fracture Charactersistics of Modern Steel Weldments and the Practical Integrity of Marine Structures Made by TMCP Steels", Proc. Int. Conf on Evaluation of Materials Performance in Severe Environments, Japanese Steel Institute, Tokyo, March. 1990.
Denys, R.M. (1990a), Wide plate testing of Weldments, Part I, II and III, Fatigue and Fracture Testing of Weldments, ASTM STP 1058, Eds. H. McHenry and J. Potter, ASTM Philadelpia, 1990, pp.157-228.
Harrison J.D. (1980), The "State-of-the-art" in Crack tip opening displacement (CTOD) testing and analysis, T.W.I. Research Report 7302.10/80/210.2
Kamath M.S. (1981), The COD Design Curve: An assessment of validity using wide plate tests. Int. Jour. of Pressure Vessels & Piping 9.
Pisarsky H.G. and Harrison J.D. (1985), Fracture Toughness Considerations for offshore shore structures in UK waters. Conf. Welding for Challenging environments, October, 85, Toronto, Canada.
Pisarski, H.G. and Walker, E.F. (1986), Wide plate testing as a back up to the CTOD approach, TWI Report for the DoE 3915/4/86.
Soete, W., and Denys, R. (1976), Fracture toughness testing of welds. Proc. of Conf. on Welding of HSLA (microalloyed) Structural Steels, Rome, Nov. ASM, ISBN 0-87170-005-0, pp.63-84.
Walker, E.F. (1986), Fracture toughness testing - Present status of Charpy V notch impact and CTOD testing, Proc. Symp. on The State of the Art in Materials Testing., K.VIV, Nov. 1986, Antwerpen.
Woodley, C.C., Burdekin, F.M. and Wells, A.A. (1964), Mild Steel for Pressure Equipment at Sub-Zero Temperatures, Brit. Weld J. 1964, 11, 3, pp.123-136.

# Materials and Their Weldability for the Power Generation Industry

Carl D. Lundin
Materials Joining Group, Materials Science and Engineering, The University of Tennessee, Knoxville, TN 37996

Abstract
The development and utility of modern materials for the power generation industry for the 1990's and beyond is discussed together with their weldability The need for additional base material research is highlighted along with the requirements for filler metal evolution.

## 1. Introduction

Future global power needs will be met by utilizing a variety of technologies. The efficiency of the particular process will depend, in part, on the materials and fabrication methods used for construction. Higher temperatures for fossil energy electric generating units will necessitate advanced alloys with superior weldability and in nuclear power generation systems it is implicit that the structural integrity be beyond question. It is without doubt that the majority of the electric power for the 1990's will be derived from either fossil fuels or nuclear energy and the materials and technologies for these systems must be optimized. Research into the alloy development and materials joining aspects covering these energy generation methods is being pursued vigorously in Europe, Japan and in the US through the Department of Energy (DOE) Fossil Energy Advanced Research & Technology Development (AR & TD) program, Electric Power Research Institute (EPRI) and the Materials Properties Council (MPC)

The incorporation of new materials into Codes governing construction of power generation facilities is a process with considerable inertia and the expectations of the Code writing bodies have risen with regard to the extent and quality of the data required for supporting the incorporation of new and improved materials into the Code documents. The changing approach by the Code bodies makes the cost of obtaining the approval quite significant and thus many fabricators and constructors will not seek such approvals.

The economic exigencies of the times have made efforts to extend the life of components cost effective. In the life extension methodology replacement of critical components is often required and these components many employ materials with more modern technology whose properties exceed those of the original construction. Further, the installation of replacement components which require less in service inspection because of fewer welds, show lower fabrication costs because of higher strength and lighter weight and/or permit the plant to be operated at higher temperature for increased efficiency is of prime concern in the power industry today.

This paper discusses some of the materials improvement which are now available for use or are on the horizon for use in the 1990's. the weldability concerns are also highlighted in the discussion of the respective materials with emphasis on potential research area to be addressed to enhance the use of these newer materials is the future.

## 2. Materials for the 1990's and their Weldability

### 2.1 Austenitic Stainless Steels for Nuclear Service

The occurrence of Intergranular Stress Corrosion Cracking (IGSCC) in the Boiler Water Reactor (BWR) systems in the US in the early 1970's lead to research which, in part, sought to improve the material response to the high temperature (288oC) oxygen containing water. In the US this research resulted in the 304 and 316 Nuclear Grade materials (304NG & 316NG) (1-5) and caused the research community to re-examine the 347 stainless steel materials which proved successful in the German BWR plants. The basic modification to the 316 chemistry was the lowering of the carbon content to <0.02% and the introduction of nitrogen in the range of 0.06 to 0.10%. The reduction in the carbon was very effective from the sensitization standpoint and the addition of nitrogen maintained the yield strength level desired and produced a fortuitously beneficial effect on sensitization by retarding the carbide growth kinetics. The TP347 grade which evolved in Europe proved to be equivalent to the 316NG in the BWR environments (6).

The weldability of these materials was addressed in a group sponsored study at The University of Tennessee (7) which showed that all of the materials had excellent hot cracking resistance when formulated with a Cr/Ni equivalent ratio providing for primary ferritic solidification and a positive ferrite potential in the wrought base material. The German TP347 was generally equal to the French version (ICL 743 Nb) and the Japanese alloy 347 LPN, however all of these newer 347 type materials made with modern commercial practices surpassed the performance of the higher carbon AISI 347. The 316NG alloy was chosen as the candidate replacement alloy for US use even though the operational characteristics of the consumables favored the TP347 material. Conservative chemistry specifications based on both weldability and corrosion resistance are shown in Table 1. This table should be used as a guide in determining the proper chemical requirements for the optimum use of these materials. It is without doubt that the TP347 material will see greater use in applications that require a high resistance to IGSCC in nuclear plant environments and the renewed interest based on positive research results for both weldability and corrosion behavior will most likely result in its more wide spread use.

Table 1. Conservative Composition Ranges for 316NG and TP347 (7).

| | C | Si | Mn | p | S | Ni | Cr | N | Mo | Nb | Co | G.S[a] | FN[b] | $Cr_{eq}/Ni_{eq}$[c] | Nb/C |
|---|---|---|---|---|---|---|---|---|---|---|---|---|---|---|---|
| 347NG | 0.03 max | 0.5 max | 1.5 -2.0 | 0.015 max | 0.015 max | 10.0 -14.0 | 16.0 -18.0 | 0.12 max | - | 0.40 max | 0.20 max | >4 | 3-5 | >1.6 | ≥8 |
| 316NG | 0.02 max | 0.5 max | 1.5 -2.0 | 0.015 max | 0.015 max | 10.0 -14.0 | 16.0 -18.0 | 0.06 -0.12 | 2.0 -3.0 | - | 0.20 max | >4 | 3-5 | >1.6 | - |

a: ASTM grain size No.
b: Ferrite Number according to Figure NB-2433.1-1 of ASME Section III (Delong Method)
c: Hammer and Svensson Method

$Cr_{eq}$ = %Cr + 1.37 x % Mo + 1.5 x % Si + 2 x % Nb + 3 x % Ti

$Ni_{eq}$ = % Ni + 0.31 x % Mn + 22 x %C + 14.2 x % N + %Cu

## 2.2 Modified 9Cr-1Mo-V (ASME T/P91) for High Temperature Service

The US DOE instituted a program in 1974 to select/develop alloys with improved behavior for Liquid Metal Fast Breeder Reactor (LMFBR) applications. Oak Ridge National Laboratory (ORNL) (8) was assigned the task of alloy development which eventually led to an improved ferritic steel for high temperature service. The 9Cr-1Mo material was selected for the operating temperature range and an alloy modification program conducted at Combustion Engineering led to the use of V and Nb for increasing the high temperature creep strength. The cost of the development program exceeded $17,000,000 in 1985 and in 1986 the modified alloy was approved by ASME. The joining behavior was also studied and no major fabrication weldability problems were encountered, save for a hardness loss in the HAZ which influences the weldment transverse creep rupture strength. However, by a judicious choice of tempering and PWHT schedules this effect can be minimized. While the steel was directed at LMFBR usage it has found its greatest utility in fossil energy plant applications.

The ORNL data which shows the superiority of the modified 9Cr-1Mo-V in the creep regime is presented in Figure 1. These data which compare the elevated temperature behavior with Standard 9Cr-1Mo and 2 1/4Cr-1Mo are self explanatory. There have been over 1000 tons of the modified material produced and the results on full commercial heats have been promising. The material is being employed in original designs for fossil fuel plants, in retrofit and replacement and in turbine rotor applications. The Japanese and the Europeans have been leader in the production of the material in tubing, piping and forgings.

Recent data obtained in a doctoral thesis investigation by F. Bruhl at the University of Graz, Austria (9) has revealed that the creep rupture strength of material produced to the ASTM P/T91 specification does not meet the published rupture strength. Figures 2 & 3 from the work of Bruhl show the differences for the base metal tests and weldment test respectively. The base

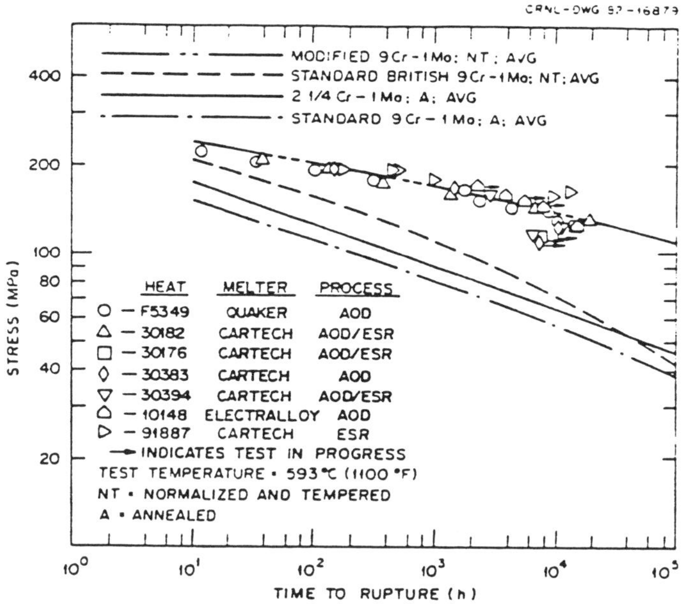


Fig.1. Stress Rupture Plot for Comparing Data on Modified 9Cr-1Mo and Other Alloys in Comparable Use (8).

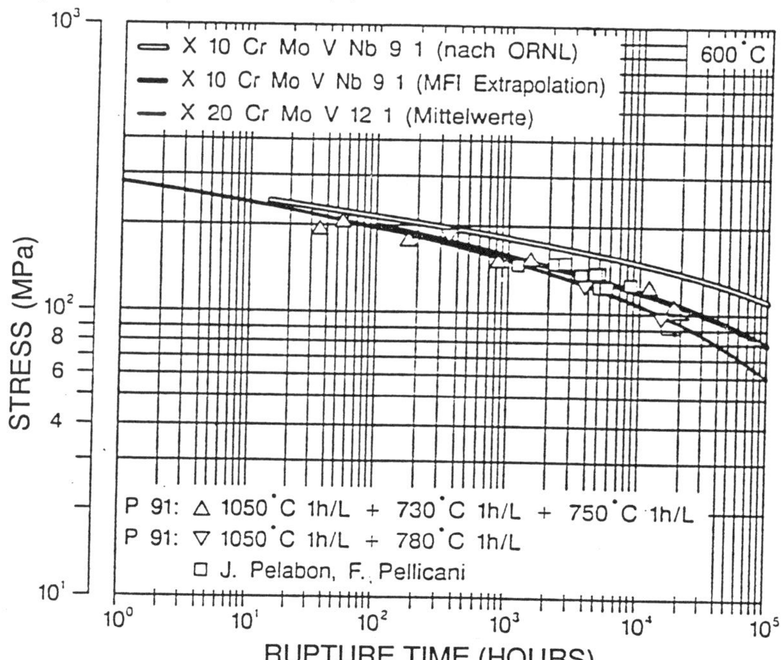


Fig.2. Creep Rupture Test Results of 9Cr-1Mo (P91) and 12Cr-1Mo Base Metal (9).

metal data of ORNL together with that of Mannesmann (MFI) on 9Cr-1Mo modified and the mean data for 12Cr-1Mo-V. The effect of heat treatment variation is also shown it is clear that the data obtained for the P91 material in this work falls below the ASME data base. The weldment data shown in Figure 3 are for traverse weldment tests incorporating the weld metal, base metal and HAZ. Data are shown for P91, Alloy 625 and Cr-Mo-W-V filler metals and again the data fall below the ORNL/ASME mean data line. All of the long time weldment test specimens ruptured in the HAZ or at the fusion line. The full significance of these data are not known at this time and more work is indicated before final sweeping conclusions can be made. It is to be noted that ASME code case N-47 was recently modified for P/T91 materials by the inclusion of stress reduction factors for weldments which are a function of time and temperature

While weldability problems have been very minimal with the modified 9Cr-1Mo-V materials some variability has been reported with toughness for weld metals deposited by the SMAW process. Further, microfissuring has also been indicated in some weld metals. However, these difficulties appear to be soluble and will probably have little effect on the future utility of the material.

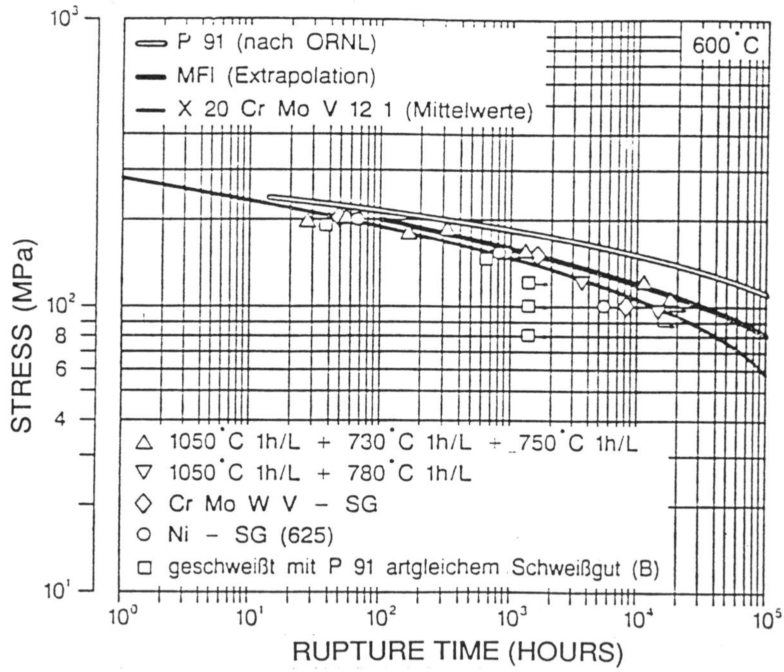


Fig.3. Creep Rupture Test Results of P91 Weldments in Comparison to Base Materials (9).

This material with its proven utility has generated world wide interest. Four papers dealing with the material were presented at the ASME Pressure Vessel and Piping Conference in Nashville Tennessee, June 1990. In these papers applications for hydrocrackers and boiler headers were discussed. Improved high temperature mechanical properties have been imparted to the material by thermo-mechanical processing and tempering. In addition,the tailoring of the chemical composition and employment of advanced ladle refining and unidirectional solidification has resulted in improved toughness and creep rupture strength. Thus, the world wide interest remains extremely high in the modified 9Cr-1Mo-V alloy system and it is to be predicted that this material will be utilized to a great extent in many future power generation applications.

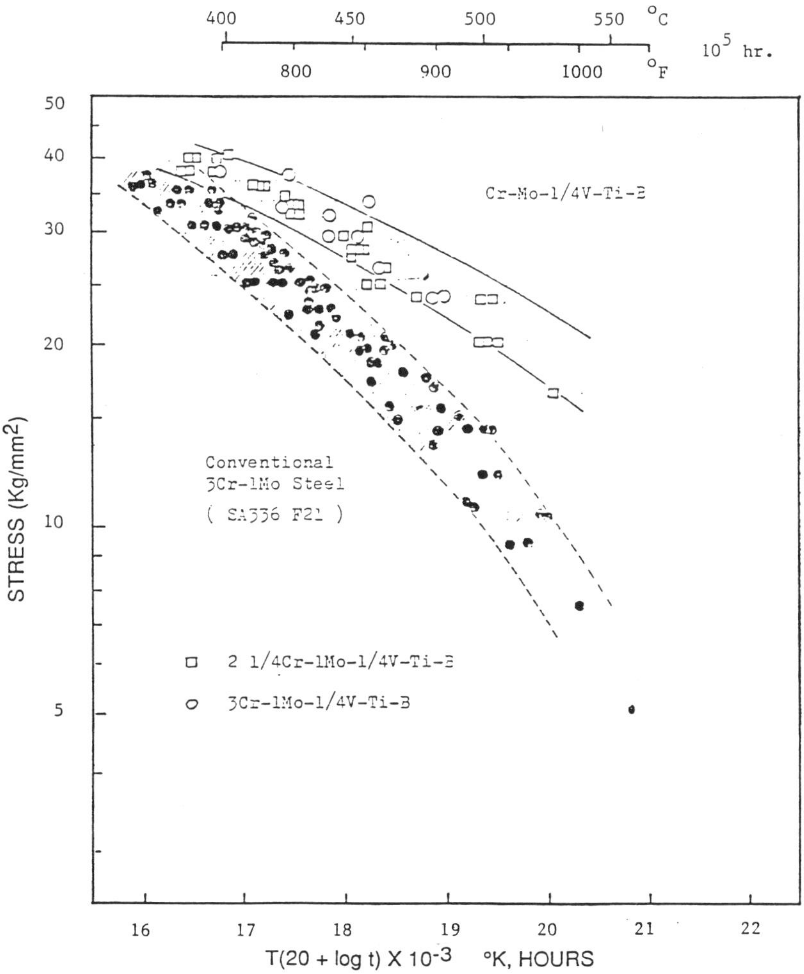


Fig.4 Creep Rupture Strength of Cr-Mo-1/4V-Ti-B Steels (11).

2.3. Advanced 2 1/4 Cr-1Mo-V Steels for Petroleum and Hydrogen Service
The Materials Properties Council (MPC) in cooperation with and sponsorship from the American Petroleum Institute (API) and a world wide group of steel makers and fabricators has fostered considerable research effort in developing Cr-Mo steels with higher strength (85-110 Ksi) and enhanced creep rupture properties for applications to 900oF (10). The majority of the steel and filler metal development was accomplished in Japan with a significant amount of testing in both air and hydrogen environments in the US. The ASME B & PV code has accepted the results and has issued Code Case 1960 which recognizes the enhanced properties of the modified materials in elevated temperature service. The early data showing the influence of the V-Ti-B additions to the 2 1/4 and 3Cr-1Mo materials is shown in Figure 4 (11,12). It is clear that the data show the enhancement of properties. Additional data for several commercial heats support these earlier trends and testing is currently underway in hydrogen atmospheres.

Filler metals were developed concurrently with the wrought material research activities and weldments have been tested which show that the joint strength can match the base metal strength with proper welding procedures (13). The fact that the filler metal development and the weldability studies, especially reheat cracking, was accomplished in such a timely manner is a direct result of MPC involvement in the program. Concurrent filler metal/weldability investigations are certainly the means by which early commercialization of an alloy development can occur.

These alloys with clearly demonstrable properties will constitute an important link in future power generation and petroleum processing development.

2.4 Dissimilar Metal Welds
The employment of transition joints (usually between Cr-Mo and 300 series austenitic stainless steels) in boiler tubing and piping components has previously presented problems in terms of premature failures in the dissimilar metal welds (DMW). The methods for improving the performance of such joints was undertaken by EPRI under the MPC management umbrella (14). The studies determined the mechanisms of failure and the means of mitigating the attendant problems.

In addition to the weld procedural aspects of the study, filler metals with ferritic and austenitic structures were evaluated and a nickel base ally filler metal was developed which was shown to improve the life of the DMW by the control of the metallurgical phase changes and resulting carbide morphologies at the DWM interface. The alloy developed was designated as HFS6. However, while the formulation was highly successful in mitigating the DMW in-service failure problems it suffered from a tendency toward microfissuring in the HAZ of overlapping weld metal passes. Continued weldability and developmental activities are necessarily to bring this very promising weld metal to full commercial utilization.

2.5 Advanced Austenitic Staidness Steels for Super Critical Steam Service
The development of advanced steam cycle (650oC-35MPa) power generation equipment is being aggressively pursued as power demands in the 1990's and beyond will have to be met with more efficient plants. The materials for use at the proposed steam temperature and pressures are being developed in Japan, Europe and in the US under DOE sponsorship, through the AR & TD Fossil Energy program. The basic tenets of alloy design involve long time strength and creep rupture properties, metallurgical stability, corrosion resistance, oxidation resistance and fabricability, which includes weldability. Austenitic alloys are prime candidates because of stability, minimization of embrittling phases and elevated temperature strength potential at reasonable cost. Strengthening improvements have been brought about by controlled precipitation of carbides, nitrides and phosphides which resist coarsening during service and through a combination of thermo-mechanical processing (cold work and grain refining and aging). The alloys span the range of the stainless steel compositions to the Alloy 800H range of alloying elements with each composition modified to induce higher temperature creep rupture strength.

To achieve optimum stability the alloys are fully austenitic so that the potential for selecting the most advantageous welding solidification mode (ferritic) is absent. Thus, the materials are susceptible to weld hot cracking in the base metal HAZ and in the weld metal if autogenously welded or welded with filler metal of matching composition. Therefore, specific considerations have to be addressed when selecting filler metals for joining these promising materials.

Fabrication procedures, especially by welding, must fall within the realm of standard practice in the fossil energy industry and be consistent with the tenets of the ASME B & PV Code. The alloys should be free from hot cracking, reheat cracking, HAZ creep rupture property deterioration, exhibit freedom from embrittlement in long time service, and be capable of being joined to other alloys without long time interface effects. These concurrent requirements place significant restraints on both base metal and filler metal development.

Hull (15) with Westinghouse, set about to develop a weldable high strength stainless steel for elevated temperature service and in doing so was one of the first researchers to show the effects of several elements on weldability (the steel devised is called Kromarc 58). Furthermore, the premise for his work was "weldability" and the associated strengthening mechanisms utilized by him still form the basis of alloy design and strengthening in the more recent alloy development programs. Unfortunately, weldability was not a significantly emphasized criteria in subsequent alloy developments. The Japanese have developed a variety of alloys which show enhanced strength in the 650oC operating range but suffer from the potential weld hot cracking problems attendant with all fully austenitic materials. Among these alloys are the Tempaloy A-1, A-2, CR30A and NAC CR35A of NKK (Nipon Kokan), HR3C and ST3Cu of Sumitomo Metals and the recently reported development by Nippon Steel of Alloy NF 709. These alloys are primarily based upon the compositions of 17-14CuMo, Standard 310 and 347H stainless steel and Inconel Alloy 800H with intentional additions of some or all of the following elements: Nb, V, Ti, N, Al Cu and B. These additions enhance strength primarily through precipitation of fine second phase particles such as carbides, nitrides or copper precipitates, the formation of which may be enhanced by prior cold work. Even with enhanced base metal compositions the current filler metals of choice (results of an informal survey of producers of filler and base materials) are based on the traditional 17-14CuMo material used in the first super critical steam-electric units at the Eddystone facilities of Philadelphia Electric constructed in the early 1960's, Inconel 625 and Inconel 617. Thus, filler metal development has lagged significantly behind the base metal developments.

While the majority of the strengthening effects induced by adding the above minor elements to stainless steels and high Ni content alloys has been known for the past 20-30 years, Oak Ridge National Laboratory researchers under the DOE Fossil Energy AR & TD program have designed several alloys which have been shown to have the highest creep rupture properties of any of the previously developed lean (low Cr) austenitic stainless and 20Cr-25Ni (800H) materials (16-19). The creep rupture properties of several commercial alloys are shown in Figure 5, together with the properties of the lean austenitic stainless steels developed at ORNL(18). All of the ORNL alloy data points fall above the strengths for existing commercial alloys. Therefore this development appears to be significant and may have great future potential.However, these alloys also suffer from a potential for HAZ hot cracking and filler metals have not been developed which match the base metal creep rupture strength.

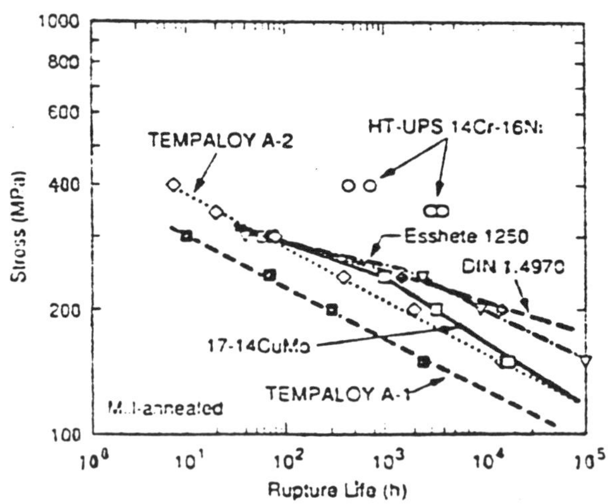


Fig.5. Creep Rupture Behavior of ORNL Lean Stainless (HT-UPS 14Cr-16Ni) at 650°C (19).

All of these alloys have a high potential to be used in future steam-electric generation plants and perhaps the lack of suitable filler metals and the potential for hot cracking will not influence certain applications. However, there is a need to address the filler metal issues as well as to enhance all fully austenitic alloys resistance to hot cracking.

## 2.6 Nitrogen Containing Austenitic Stainless Steels

The nitrogen containing austenitic stainless steels fall into three ranges of nitrogen content (0.05-0.1, 0.1-0.4 and >0.4, in wt. %). Standard grades with nitrogen in the range of 0.06-0.1% are fast becoming popular because of the strengthening effect benefit if carbon content is lowered and for enhanced corrosion pitting and IGSCC resistance. This level of nitrogen is easily maintained in the weld metal without special procedures and weldability is not sacrificed. The intermediate range of nitrogen in stainless steels is strongly affected by composition especially the Mn and Cr content and the Nitronic series of alloys have been available for many years. However, filler metal selection is still an area of concern depending on the application and Inconel 82, Alloy 625, 625+ and others have been utilized. However, considerably more effort is necessary to make informed choices based upon specific applications. For the fully austenitic grades the hot cracking sensitivity problem is similar to that of other austenitic stainless steel and dilution related problems become a difficulty with many process/filler metal combinations. Welding in nitrogen containing atmospheres or at hyperbaric pressures with nitrogen can effectively maintain a high level of nitrogen in the weld fusion zone and this avenue of effort should be explored for optimum property matching.

Nitrogen containing steels with nitrogen contents from 0.4 to greater than 1.3% represent a class of material that can exhibit high strength and toughness especially in the cold worked condition. However, the joining of these materials to obtain matching joint properties has not been adequately addressed and there are essentially no filler metals with similar metallurgical characteristics that can match the mechanical properties. Furthermore, the HAZ response of these materials has not been documented and ways of mitigating the reduction in strength due to recrystallization have not been addressed. While the base metal properties appear to be very interesting and useful, there is little known about weldability and thus, there is a strong need to develop filler metals and to determine the weldability relationships for these newer high nitrogen stainless steels.

## 2.7 High Molybdenum and Molybdenum-Nitrogen Austenitic Stainless Steels

In the past 10-12 years a variety of high molybdenum/nitrogen bearing austenitic stainless steels was introduced to provide an improved resistance to pitting resistance, especially in chloride containing environments, and for crevice and stress corrosion cracking mitigation. The high Mo content is primarily responsible for the improved corrosion behavior with the addition of N made to improve strength and simultaneously enhance the resistance to sensitization through control of the carbide precipitation rate and the reduction in sigma formation potential (20). Representative of some of these steels are; AL-6X, AL-6XN and 254 SMO.

Some general weldability problems became apparent early on with the high Mo/N materials and these problems are just now being understood and resolved: HAZ precipitation of sigma phase at high heat inputs as modified by the presence of N, degraded corrosion resistance in autogenous welds or in the "unmixed" region of filler metal added welds and hot cracking related to fusion zone chemistry and dilution. While many of these problems have been addressed by using low heat input, over-matching with a filler metal higher in Mo, solution heat treatment and controlling dilution in the fusion zone, the initial difficulties did, and in some applications still do, present complications. Had a significant weldability program been initiated during the alloy development stage many of these difficulties could have been avoided in production weldments.

## 3. Closure

From the foregoing it is evident that there are several new materials whose base line properties and metallurgy have been investigated to the extent that they are candidates for use in power generation equipment for the 1990's and beyond. However, many of the developments have not been brought full circle, in that, the ability to effectively join the materials has not been fully defined. Further development especially in the mechanical behavior of weldments and in the development of filler metals is in order.

## References


1. Hanneman, R.E., Danko, J.C., Giannuzzi, A.J. and Riccardella, P.C.,Cause of Cracking-Evaluation of Performance Margin Between Different Diameter Type 304 Stainless Steel Piping Systems", EPRI NP-944, Dec. (1978)
2. U.S. Nuclear Regulatory Commission Pipe Crack Task Group, Investigation and Evaluation of Stress-Corrosion Cracking in Piping of Boiling Water Reactor Plants", NUREG -1061, Mar. (1984).
3. Alexander, et al., "Alternative Alloys for BWR Piping Applications", EPRI NP-2176, Oct. (1982).
4. Kobayashi, et al., "Stress Corrosion Cracking Resistance, Mechanical Properties and Weldability of Nuclear Grade 316 Stainless Steel", EPRI WS-79-176, May (1980).
5. Lundin, C.D., Lee, C.H., Menon, R., and Stansbury, E.E., "Sensitization of Austenitic Stainless Steels: Effect of Welding Variables and HAZ Sensitization of AISI 304 and HAZ Behavior of BWR Alternative Alloys 316NG and 347", Welding Research Council Bulletin No. 319, Nov. (1986).
6. Shack, W.J., Kassner, T.F., Maiya, P.S., Park, J.Y., and Ruther, W.E., "Environmentally Assisted Cracking in Light Water Reactors" NUREG/CR-4667 Vol IV-ANL-87-41 Dec. (1978)

7. Lundin, C.D., Lee, C.H. and Qiao, C.Y.P., "Group Sponsored Study-Weldability and Hot Ductility of Nuclear Grade Austenitic Stainless Steels", Final Report, Welding Research and Engineering Group, The University of Tennessee, Knoxville, TN, Dec. (1988).
8. Steffano, J.R.,et al. "Summary of Modified 9Cr-1Mo Steel Development Program, 1975-1985" ORNL 6303, Oct. (1986).
9. Bruhl, F., "Vrhalten des 9%-Chromstahles X10CrMoVNb91 und seiner Schweisverbindungen im Kurz- und Langzetiversuch" PhD Thesis, University of Graz, Austria, May (1989).
10. Prager, M., Cox,T.B., Wada, J.,"Enhanced and Modified Materials For Higher Temperatures and Pressures", Proceedings of the International Conference on Interaction of Steels With Hydrogen in Petroleum Industry Pressure Vessel Service, Materials Properties Council, N.Y., N.Y., Mar. (1989).
11. MPC Meeting Minutes, HPV-7, May (1983)
12. Japan Steel Works Report MRH 83-3-18, Mar. (1983).
13. Yamaura, T., et al.,"Mechanical Properties of 2-1/4Cr-1Mo-V SA-Welds by a Single Pass Per Layer Welding Method" Kawasaki Steel Corporation, October 1989, MPC, HPV-47 Oct. (1989).
14.EPRI, "Dissimilar-Weld Failure Analysis and Development Program" EPRI Project 1874, Volumes 1-8, (1989).
15 Hull,F.C. "A High Strength Weldable Stainless Steel For Elevated Temperature Service" ASTM STP 369, Advances in the Technology of Stainless Steels, 1965.
10. Swindeman, R. W., Goodwin, G.M., Maziasz, P.J., Judkins, R.R., and DeVan, J.H., "Alloy Design Criteria and Evaluation Methods for Advanced Austenitic Alloys in Steam Service" ORNL-6274, May. (1986)
16. Swindeman, R. W., Goodwin, G.M. and Maziasz, P. J. "Procurement and Screening test Data for Advanced Austenitic Alloys for 650oC Steam Service (Part 1, 14 Cr-16 Ni and 20Cr-30Ni-Fe Alloys)", ORNL/TM-10206/P1, Apr. (1987).
17. Carolan, R.A., Li, C.Y. Maziasz, R.W., and Todd, J.A., "The Strengthening Mechanisms in Modified AISI 316 Stainless Steels", ORNL/SUB85-27488/03, Apr. (1987).
18. Black, R.D., Carolan, R.D., Li, C.Y., Swindeman, R.W. and Maziasz, P.J. "Effect of Thermal-Mechanical Pretreatments on the Flow Strength of a Copper Modified, Stabilized 316 Stainless Steel", ORNL/SUB/85-27488/03, Jan. (1989)
19. Maziasz, P.J. "Developing an Austenitic Stainless Steel for Improved Performance in Advanced Fossil Power Facilities", Journal of Metals, July (1989)
20. Rabensteiner, G. " The Welding of Fully Austenitic Stainless Steels With High Molybdenum Contents", Welding In The World, Vol. 27, No. 1/2 (1989).

# TRENDS IN DESIGN AND FABRICATION OF SHIP STRUCTURES

H. PETERSHAGEN
Institute of Naval Architecture, University of Hamburg

Abstract
Modern ocean-going vessels are complicated tailor-made systems with a present and probably future trend towards specialisation and higher economy. Ship structures are developed according to the requirements imposed by systems design and with a view to cost-saving production. The finite element method dominates in the strength analysis. While large models are used for the overall analyses, design and analysis of details needs further development, especially with regard to fatigue assessment. Low carbon steels are generally used for ship structures, but high strength steels, light alloys and reinforced plastics are applied in certain cases as small high-speed crafts.

Current production methods are modified to reduce production costs. Plasma cutting, laser cutting and welding and robot welding are mentioned as examples for introduction of new processes into the fabrication of ship structures. Computer application can be considered as an integrating factor for the whole process of ship design and fabrication. Present state and trends are briefly described.
Key words: Ship structure, design, materials, fabrication methods, computer application.

## 1. Introduction

In reporting about present state and possible developments in shipbuilding under the auspices of IIW, one may feel on a rather safe ground with the first, but quite uncertain with regard to the second aspect.

In fact, the development of ships and shipbuilding during the last three to four decades has been as rapid as it has been unexpected in many cases. Design of new types of vessels, introduction of computer technology into design and fabrication and application of modern welding and cutting processes may be mentioned for illustration.

Nevertheless, an attempt has been made here not only to describe different aspects of the present situation but also, with all reservation, give examples for possible future developments.

## 2. Ship design

Modern ocean-going vessels are complicated systems. They have to meet a variety of demands, which may be summarized under the headings safety, economy and hability. These demands play an important role under the different aspects of development discussed in this paper.

Demand of economical transportation as well as the variety of goods to be transported has led to an increasing specialisation of ships. The heavy cargo/container vessel shown in fig. 1 may serve as an example. This tendency for specialisation is combined with the building

Fig. 1 Heavy cargo / container vessel

of small series or even single units. Thus ships differ from aircrafts or motor cars as in most cases they are not only complicated but also tailor-made systems.

It may be expected that the trend towards specialisation and higher economy will continue. For example container vessels with increased height and without hatch covers are presently discussed (fig. 2). It is immediately evident that safety against sinking and capsizing caused by sea and rain water entering into the holds would be a key problem in this type of ships.

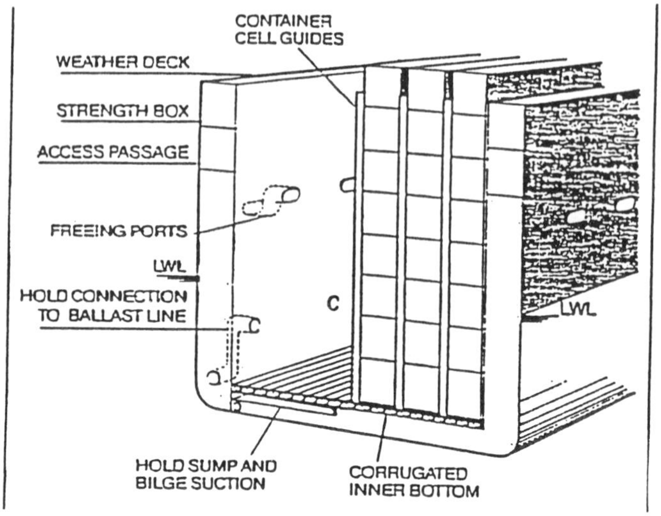


Fig. 2 Hatchless container vessel

A special branch in the development of ships are high-speed small crafts, e.g. surface effect ships (SES), catamarans and SWATHS. Fig. 3 shows an SES developed by Blohm and Voss AG in Hamburg. They are in the present context of interest especially with regard to the materials used.

Fig. 3 Surface effect ship

## 3. Structural design and strength analysis

Roughly speaking ship structures are built up from stiffened panels. One distinguishes between the hull girder, which is treated as a beam subjected to bending and torsion, and local structures. Of course, many structures act as parts of the hull and local structures as well.

The present and future task for the designer is to develop structures which meet the requirements of cost-saving production methods and those imposed by systems design while maintaining sufficient structural safety.

Characteristic examples for the first task can be taken from the field of design for welding robots. Fig. 4 shows the penetration of longitudinal stiffeners through transverse webs, a very frequent element in ship structures.

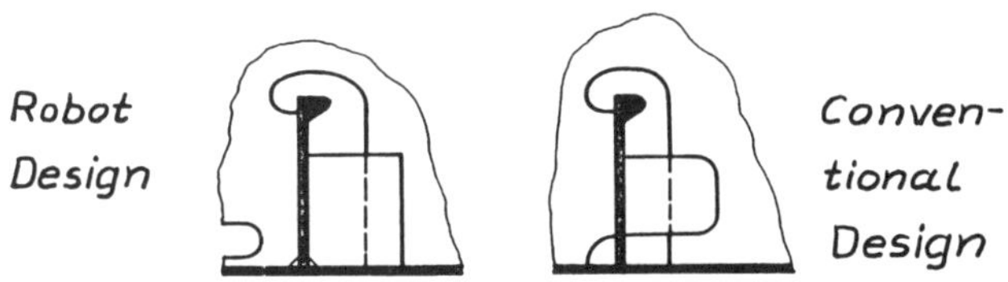


Fig. 4 Connection longitudinal stiffener - transverse web

A double bottom structure designed for extensive application of automated welding processes is described in Begler and Kahle (1989). The number of transverse floors has been reduced to one per section, thus making the structure accessible from the ends during the assembly (fig. 5). Strength analysis and weight assessment showed

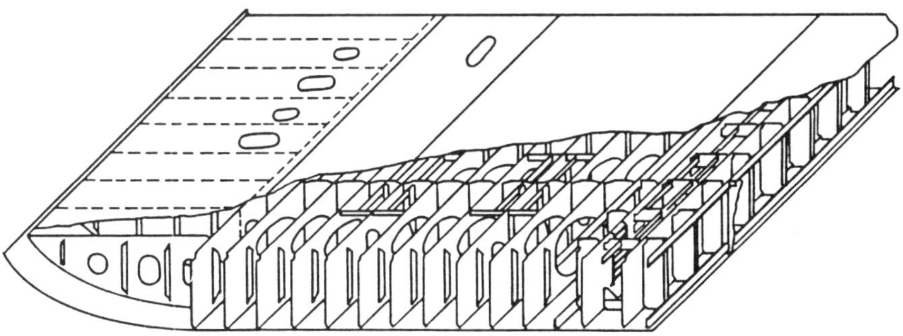

Fig. 5 Double bottom structure

no disadvantages compared with a conventional design. The advanced design is of course applicable to other closed box-type structures as twin shells and bulkheads.

An example for a structure developed with regard to system design are transverse girders with large web openings (fig. 6 from Höft (1989)). Typical areas for application are accomodation decks of passenger vessels and ferries. Air ducts, pipes and cables have to be arranged underneath the girders or penetrating their web. A further requirement is a web height providing a sufficiently large section moment of inertia of the girder and hence a sufficiently high natural vibration frequency of the deck.

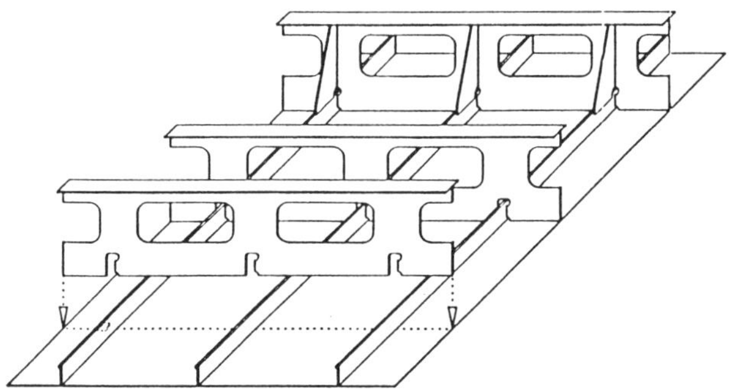

Fig. 6 Girder with large web opening

On the other hand, the full heigth from deck to deck has to be kept as low as possible with regard to the overall height and the stability against capsizing of the vessel. Girders with large web openings for ducts, pipes and cables have been shown to be an excellent solution with regard to all these requirements and are now widely used.

Finite element methods (FEM) are used in the strength analysis of ship structures to a large extent. The whole hull is often modelled in case of sophisticated, highly loaded ships such as open-type vessels. Fig. 7 shows the model of a finite element calculation carried out by the German Lloyd for the ship shown in fig. 1. A gap to be filled up is the design of structural details, which is often still done empirically. Many ship structures are predominantly fatigue loaded, and it is well-known that fatigue properties are to a large extent a matter of detail design and workmanship. Fig. 8 shows an example, where a fatigue crack was formed in the hatch coaming of an open-type vessel. At such locations the rails for longitudinal movement of the hatch covers are interrupted in order to arrange a hydraulic operated device for lowering the cover. In the lowered position the sealing between hatch cover and coaming becomes effective provi-

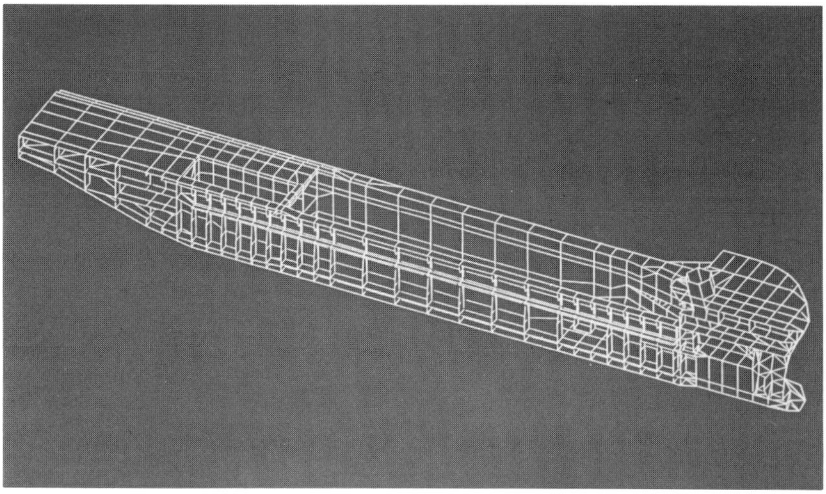

Fig. 7 Finite element model of an open type vessel

Fig. 8 Fatigue crack in hatch coaming

ding watertightness of the hatch cover. On the other hand, the continuous hatch coaming acts as a part of the load-carrying hull and the interruption forms a very sharp notch leading to fatigue cracking. At present, research programs are carried out to extend the use of fatigue assessment methods in the design and strength analysis of ship structures. This development can be considered as a part of the process of introducing rational design criteria into the strength assessment of ship structures, which could be observed during the last three to four decades. The introduction of reliability based methods may well be a further step in this direction in the future.

## 4. Materials

For several reasons low carbon steels with yield points up to 355 $N/mm^2$ are generally used for ship structures. In certain cases, alloyed steels are applied, where low temperature application or chemical resistance are required. However, further development can be predicted in certain areas.

As an example the development of small high speed crafts as surface effect ships (SES), fast catamarans and SWATH's (small waterplane twin hull ships) may serve. In these vehicles with speeds in the range between 30 and 100 knots, weight saving is essential. Besides a sufficient knowledge of the forces acting on the structure in service and the application of advanced strength analysis methods material selection is an important design aspect Pittaluga (1989).

Materials used up to now are high and ultra high strength steels, aluminum alloys and glass reinforced plastics. Other light materials as titanium alloys, carbon reinforced plastics and other high strength fiber reinforced plastics seem to be too costly at present but may become interesting in the future.

Aspects to be considered are resistance against corrosion in a marine environment, resistance to fire, acoustical characteristics and, last not least, the technology of connections.

Stiffened panels in light weight structures have thin plates and small stiffener spacings. Special profiles combining stiffener and plating and connected by butt welds can be cost-effective solutions.

The fine structural texture implies an increase in weld length, and together with the increased notch sensitivity of high-strength materials and high impact loads in service, a higher demand on weld quality. All this leads to higher requirements regarding quality control during fabrication and to shorter inspection intervals during service. The latter, however, is no problem for advanced vehicles operating on short trips and frequently drydocked for the purpose of cleaning the hull and inspecting the propulsion equipment. Typical examples for use of high strength/low weight structures are wings of hydrofoils and struts of SWATH crafts.

Sandwich elements are increasingly used as building elements for fast passenger vessels. However, their acoustical properties can be very poor. If the acoustical characteristic could be predicted from material parameters, plate geometry and frequency range, then sandwich plates could be optimized with regard to weight, strength and noise protection.

## 5. Fabrication

Cutting and welding form an important part in the fabrication of ship structures, the whole fabrication process being at present characterized by a still relatively low level of mechanisation (Seyffahrth (1989) . One major reason for this is limited accuracy during the different production stages, especially in curved parts. Another reason, connected with the individual design of ships, is the low degree of standardisation in ship structures.

Introduction of new production methods are a suitable means to improve this situation. The processes introduced are often already known and applied in other fields, but have to be adapted to meet the special requirements of shipbuilding.

On the other hand, changing types of vessels and structures and the requirement of cost-saving production lead to modifications in the application of current methods and will certainly continue to do so.

### 5.1 Developments in the application of current methods

Three examples will be given for a development towards a more rational production. A careful investigation of strength properties is often indispensable in such cases.

Fillet welds between plate and stiffener are often oversized due to production requirements. Intermittent welding is a conventional means for cost reduction. However, this type of weld is not applicable in connection with mechanized processes like SAW and has, moreover, the disadvantage of poor fatigue properties. It has been shown by experimental investigation von Selle (1987) that one-side continuous fillet welding avoids both disadvantages while maintaining the cost benefit. It can, in the present stage, not be applied in areas where special corrosion problems exist, e.g. in tanks. It appears possible, however, to develop forms of edge preparation which permit full corrosion protection and hence unrestricted application of one-side fillet welding (fig. 9).

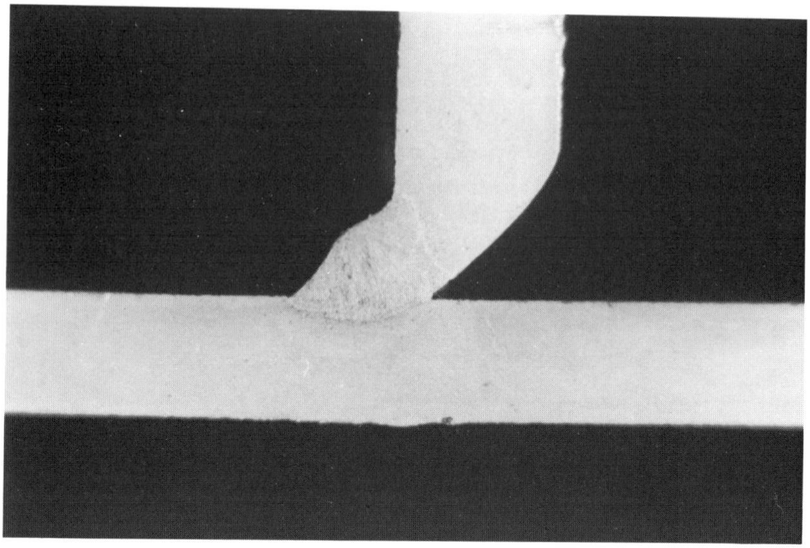

Fig. 9 One-side fillet weld

Downhand-welding of vertical fillet welds cuts the costs by a factor of about two against the convential method of upwards welding. However, because of danger of defects like slag inclusions and lack of fusion downhand fillet welding was not accepted by Classification Societies for load-carrying members of ship structures. It has recently been shown Petershagen (1989) that with the stage of development reached especially with regard to consumables and when carried out by trained welders, downhand fillet welds are sufficiently defect-free and exhibit adequate static and fatigue strength. This has led to a widened application of the cost-saving procedure also for load-carrying structures.

Partly-penetrating butt welds (fig. 10) could be used in cases, where stresses parallel to the weld dominate Petershagen (1989). This is for instance the case with longitudinal seams in side and deck plating of the ship's hull. Cost saving is achieved by a reduced volume of the weld groove in case of hand welding or by an extension of plasma-cut I-grooves to higher plate thicknesses.

Fig. 10 Partly-penetrating butt weld

5.2 Introduction of new processes into the fabrication of ship structures

Cutting is one of the important processes in the fabrication of ship structures and is often a bottle-neck in the production line. Cutting at an increased speed is therefore highly interesting for shipyards.

This can be achieved especially in the range of small plate thicknesses by applying plasma cutting instead of conventional flame cutting, see fig.11 from Decker (1989).

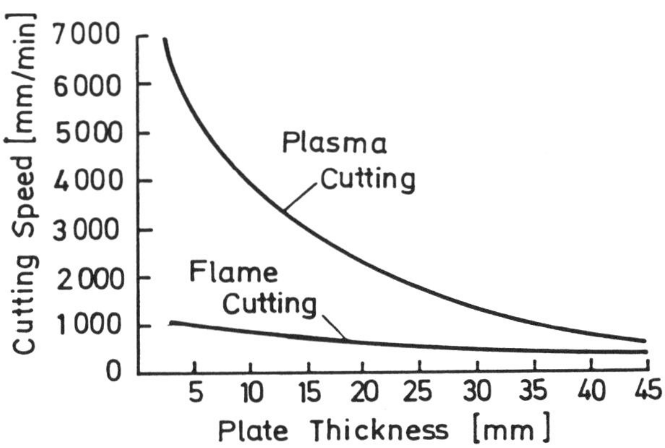


Fig. 11 Cutting speed flame vs. plasma cutting

Other advantages of the plasma cutting process are high accuracy and small deterioration due to low heat input. Noise, radiation and toxic fumes have led to the development of the water-injection-process, where cutting is performed under water. In order to avoid difficulties in the production process the further development goes to plasma cutting at the water surface. Under water plasma cut edges have mechanical properties, especially fatigue properties, which are not inferior to those of flame cut edges Paetzold (1989).

Further development in plasma cutting is directed towards quality improvement and a more flexible application. This includes optimisation of cutting parameters, modification of power units for computer control and supervision of process parameters. Control of gas emission, noise level and radiation has to be ensured during all stages of development.

Laser cutting and -welding is presently being introduced in several yards, e.g. in the United States, Great Britain and Sweden. Its major merits are extremely low heat input and high accuracy Bakowsky (1989). They can lead to a higher degree of mechanization by avoiding manual fairing during the different stages of assembly Seyffahrth (1989). Laser application is presently of interest especially for smaller plate thicknesses. The range for economical application can be increased by the

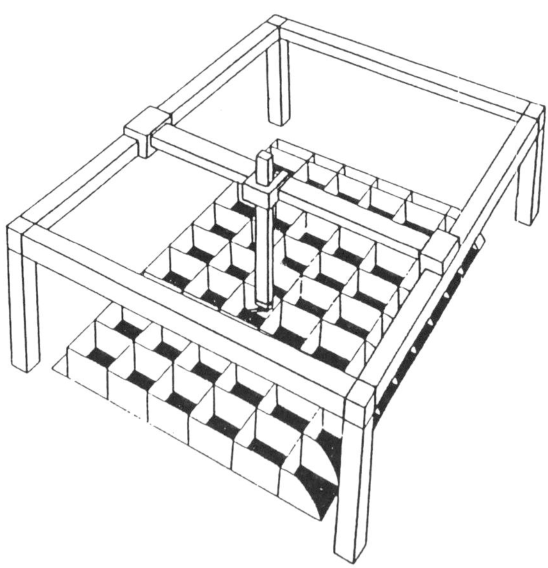

Fig. 12  Portal frame system for robot welding

development of smaller laser generators with higher output. This will permit the installation of laser units on movable portal frames for cutting and welding of large parts of ship structures.

Special attention deserves the introduction of welding robots into the fabrication of ship structures. Problems involved are individual design, dimensions of sections and environmental conditions.

Robot systems have especially been developed for ship structures accessible from the top. Three types of systems can be distinguished according to the dimensions covered:

Large portal frame systems can cover whole sections (fig. 12). A system of this type has been installed in the Wärtsilä Shipyard in Finland.

Smaller stationary systems can operate in parts of sections.

Small portable systems (fig. 13) offer a maximum of flexibility, especially in curved sections. They require, however, more additional manual operations than stationary systems. This type of systems has especially attracted attention in Japanese shipyards.

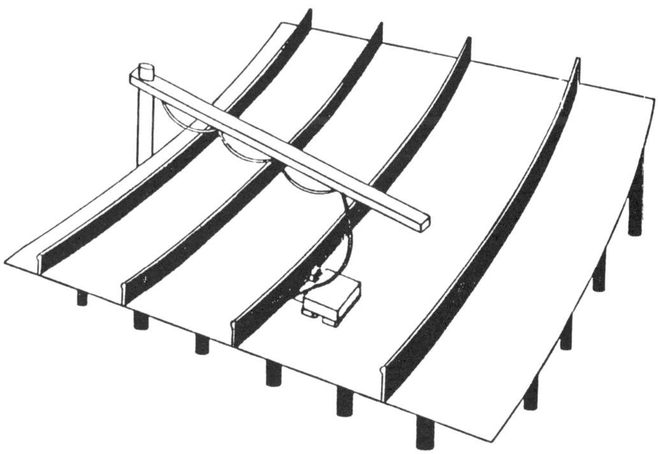

Fig. 13 Portable welding robot

A system of the second type is presently being developed in the Federal Republic of Germany Warnecke and Utner(1989). An analysis showed that welding and cutting requires more than 40% of production costs in the stage of pre-assembly and assembly of sections. A further result of the analysis was that nearly 50% of the sections, which have to be stationary assembled, consist of "boxes" with dimensions up to 3,3 x 2,5 x 2,4 m³. Double bottoms form a considerable part of this type of structure in each vessel. Double bottom boxes contain a comparatively high number of parts and welded connections with similar geometries (fig. 14). The total amount of work to be expected per year justified the development of a robot system especially suitable for welding double bottom boxes (fig. 15).

The production tolerances to be expected were investigated with a view to a proper selection of sensors.

The robot has three main rotation main and three hand axes. These six axes provide the necessary movability for the welding process. The portal frame can be adjusted to the dimensions of the actual box. The robot moves within this portal frame in longitudinal and vertical direction.

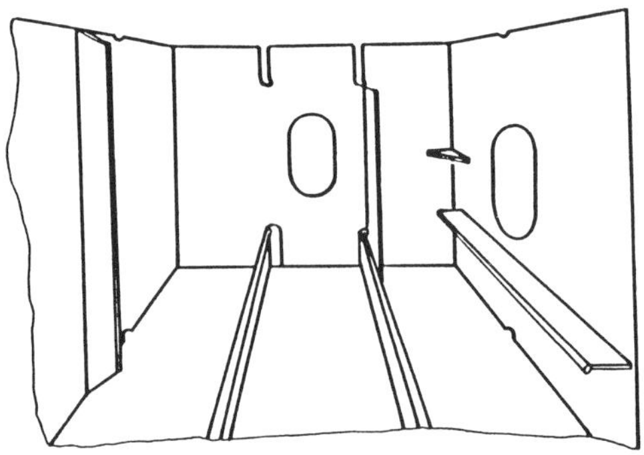

Fig. 14 Double bottom box

Fig. 15 Robot system for double bottom welding

Available "teach-in" programming systems turned out to be ineffective in view of the small number of identical boxes. However, within each box wall exists a number of similar weld patterns, which can be defined as "macros". Macro programming thus turned out to be an economical method for the present task.

## 6. Computer application

Computer application can be considered as an integrating factor for all stages of ship design and fabrication and its brief discussion in general terms appears appropriate at the end of this paper.

According to Nowacki (1989) a complete computer integrated manufacturing (CIM) system for shipbuilding would comprise activities as shown in Fig. 16. They can be subdivided into three main areas. While former activities have been directed towards isolated solutions in one of these areas, recent developments are increasingly leading to modular software architectures with open interfaces. The "product model ship" comprises the complete data base. It consists of all necessary geometrical, physical, technological but also management informations. The conventional model includes specifications, drawings, tables etc. .

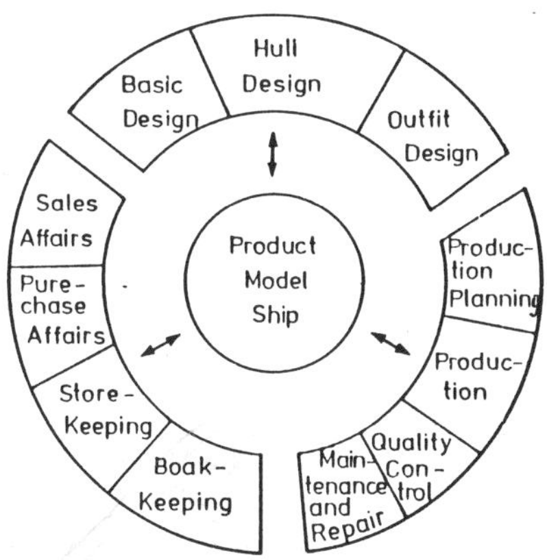


Fig. 16 Product model "ship"

The development goes towards a digital description, from which necessary other types of information can be produced. The product model ship is generated mainly during the design phase. This process can be considered as an interaction between many steps in model generation, calculation and control. The system must permit to perform each step at an arbitrary moment if the product model is complete enough to do so. The product model can be divided into submodels for hull form, space subdivision, internal hull structure, strength analysis and outfit.

In the present stage communication and exchange of data between partial systems is a key problem. For instance, models of the internal hull structure may serve to describe the real structure, e.g. for the preparation of working drawings, or to derive models for strength analysis, e.g. according to FEM. Most present CAD systems are generating descriptions of real structures. The data for the strength analysis model must be generated by special software from these descriptions or even separately.

Another problem already touched is data generation for mechanized or robotized cutting and welding from the data bases for hull form and internal hull structure. One step in this respect is the data generation for NC cutting machines including additional attributes as edge bevel information.

Important fields for the future development are described in Koch (1988). One is the integration of different systems and compilation of data in the product model ship. A standardized data description would also promote the communication between the systems.

A standardisation of functions used in different software programs could lead to libraries of such functions accessible to all software programs used.

Finally, portable computer service systems would make the use of software in a large variety of hardware products and configurations possible.

## References

Bakowsky, L. (1989) Cutting with High Power $CO_2$-Laser. Jahrbuch der Schiffbautechnischen Gesellschaft.

Begler, W. and Kahle, A. (1989) Möglichkeiten und Probleme bei der Gestaltung von automatisierungsgerechten Doppelbodenkonstruktionen (Design of double bottom structures for automation). Schiffbauforschung Special Issue. Vol. 5, 9-24.

Decker, K. (1989) Plasma fusion cutting in shipbuilding. Jahrbuch der Schiffbautechnischen Gesellschaft.

Höft, H.G. (1989) Schiffbauliche Biegeträger mit großen Ausschnitten (Girders with large openings in ship structures) Institut für Schiffbau der Universität Hamburg, Dissertation.

Koch, Th. (1988) CAD Applications in Structural Design. Jahrbuch der Schiffbautechnischen Gesellschaft, Vol. 82, 365-373.

Nowacki, H. (1988) System Architecture and Interfaces for the Ship Design Process, Jahrbuch der Schiffbautechnischen Gesellschaft, Vol. 82,343-351.
Paetzold, H. (1989) Contribution to the quality assurance of under water plasma cut edges in the ship structure. Forschungszentrum des Deutschen Schiffbaus, Report No. 211.
Petershagen, H. (1989) Festigkeitsuntersuchungen an fallnahtgeschweißten Doppel-T-Stößen aus Schiffbaustahl (Strength of downhand welded cruciform joints made of shipbuilding steel). Welding & Cutting, 41 No. 6, 285-289
Petershagen, H. (1989) Fatigue strength of non-penetrating butt welds in DH36 shipbuilding steel. IIW-Document, XIII-1315-89.
Pittaluga, A. (1989) Material selection and lightweight structural concepts and Hayman et al. (1989) Structural design aspects of fast catamarans and surface effect ships, 13th WEGEMT School: Design Techniques for Advanced Marine Vehicles and High-Speed Displacement Ships, Delft, 23rd October - 1st November 1989.
Von Selle, H. (1987) Investigastion of one side welded fillet welds for ship structures. Forschungszentrum des Deutschen Schiffbaus, Report 190.
Seyffarth, P. (1989) Rationalisierungsmöglichkeiten durch Laserstrahltechnologie für das Trennen und Schweißen im Schiffskörperbau (Possibilities for the rationalization of ship hull production by application of laser beam cutting and welding). Schiffbauforschung, Vol. 28, No.1,22-30.
Warnecke, H.J. and Utner, W. (1989) Development of an Industrial Robot for Welding in Shipbuilding. Jahrbuch der Schiffbautechnischen Gesellschaft.

# MATERIALS SCIENCE STUDY OF SPECIMENS OF PERMANENT JOINTS AND COATINGS MADE IN SPACE

B.E. PATON and V.F. LAPCHINSKII
E.O.Paton Electric Welding Institute of the Ukr. SSR Academy of Sciences Kiev, USSR

Abstract
Specific space conditions (zero - G, pressure and composition of ambient atmosphere, large temperature differences) result in the fact that permanent joints and coatings greatly differ in properties and structure from those made on the Earth.

These differences are less apparent in welding, when time of the molten pool elementary volume existence is short. In brazing they are somewhat larger. But the greatest differences are observed in deposition of coatings by the method of evaporation and condensation of materials out of liquid phase, because a material evaporated is first in the liquid state for a long time, then it goes over into the vapour phase and is condensed on substrates avoiding the liquid phase.

The paper considers the specific chemical, structural and phase features of the materials subjected to remelting, evaporation and condensation in space in the process of permanent joint formation. It is shown that these features can be either of positive or of negative character. Some ways to control the structure and the properties of permanent joints and coatings made in space are indicated.
Key words: Space, Welding, Brazing, Coating, Adhesion, Evaporation, Convection, Degradation.

Specific space conditions (zero-G, pressure and composition of ambient atmosphere, large temperature differences) [1] result in the fact that permanent joints and coatings made and serviced in space greatly differ in properties and structure from those made on the Earth. Here, one can distingwish two causes of the observed differences.

The first one is directly related to the conditions of the processes performed in space. The second is assiciated with the service conditions of permanent joints thus produced.

The process conditions are least pronounced in welding. Material being melted is, as a rule, well wetted by its own melt; therefore, no lacks of fusion in welds made in space are observed. In welding of alloys containing volatile components (e.g. magnesium, zinc) one can often observe the higher porosity [2, 3].

All this predetermined the development of special techiques allowing to eliminate occurrences of defects. As a rule, macro- and microstructures of welds made on the Earth and in space have negligible differences between each other. It applies to the weld zone as well. As for the majority of structural materials (aluminium and titanium alloys, stainless and heat-resistant steels), the structure of welds in them differs only in the finer grain boundaries and the slightly lower content of $\alpha$ -phase. Sometimes, electron microscopy indicates some differences in dispersion of $\alpha$ -phase and in the quantity of dislocations. In specimens made in space $\alpha$ -phase is usually more dispersed and dislocation density is higher. Apparently, these differences are related to the high cooling rate and the short dwell time of welding pool in molten state.

Peculiarities of space vacuum result in that metal of welds made in space is almost hydrogen-free and is somewhat enriched in oxygen. Thus, for example, in manual electron beam welding of titanium alloy the mass fraction of hydrogen in base metal was 0.0056...0.0072%, while in welds it was 0.0010...0.0015 %. Accordingly, the oxygen content of base metal is 0.058...0.061% and of weld - 0.071...0.084% [4].

The space conditions exert somewhat higher effect on the brazing processes. In brazing the relatively low-temperature filler alloys are in molten state for much longer time and have the distinct constant interface with solid-phase surfaces being joined. Under the active action of surface-tension forces thermal-capillary and concentration-capillary convection has enough time to develop during razing. Therefore, the structure of razed joints is characterized by much higher microgeterogeneity. The structure of the brazed joint zone greatly depends on the type of materials being joined; hence, it is difficult to find some general regularities. Only, the structural heterogeneity of different zones of a brazed joint much higher than on the Earth can be observed. It can be accounted for by large temperature gradients due to the different cooling conditions. Practically no lacks of penetration in brazing are observed in space.

The largest differences occur in deposition of coatings by the method of evaporation and condensation of materials out of liquid phase. In this case a material evaporated is first in the liquid state for quite a long time, then it goes over into the vapour phase and only afterwards, avoiding the liquid phase, it is condensed on substrates. Thus, at all the stages of this process one has to do with the two-phase systems, i.e. solid-liquid (in crucible), liquid-vapour (at the surface evaporated) and vapour-solid (in the substrate zone) [1]. At the first stage it is necessary to provide the reliable confinement of molten metal in the open crucible out of which evaporation occurs. It can be done only in the crucibles of a material which is wetted by a liquid evaporated, otherwise the liquid leaves the nonwettable crucibles in the form of spherical drops.

The reliable confinement of a melt evaporated is attained due to the special form of the preset temperature gradient created at the crucible inside surface. Here, the liquid is pulled to the coldest zone by mass transfer induced by thermal-capillary convection. When multicomponent systems are evaporated, one should take into account the concentration-capillary convection as well. The same processes prevail also at the second stage, i.e. separation between the liquid and vapour phases. Here, it is important to provide the preset stoichiometric composition of the vapour flow and to eliminate the drop liquid phase in it. Under the Earth conditions it is usually provided by the difference in densities between the liquid and the vapour. In space the only way to reliably separate the phases is to preset mass transfer determined by the said types of convection. As for stoichiometry and the required distribution of the vapour density in the flow, these parameters are also provided by the certain crucible shape and the preset mass transfer.

The fact, that convection caused by the density difference is absent in zero-G, allows to flexibly adjust the stoichiometric composition of condensates. It applies especially to multicomponent systems. During evaporation under the Earth conditions the similar effect is rather difficult to obtain. In space, despite the fact that the evaporation technique is much more sophisticated, it is possible to obtain coatings characterized by the properties unachievable under the Earth conditions [5].

During operation the structure and the properties of permanent joints, just as of structural materials proper, are subjected to irreversible changes under the effect of the space environment. One of the main causes of these changes, in addition to vacuum, micrometeorities, ionizing radiation and atomic oxygen, is the specific thermal cycling. It is caused by systematic crossing the terminator line by spacecrafts operating in the near-Earth orbit. Here, the flux of full solar radiation varies from maximum (about 1400 $W/m^2$) to about zero. The similar cyclic variations affect the temperature of the space craft units and permanent joints.

Table 1 gives the diagram of complex effect of the basic factors leading to gradual degradation of the properties of structural materials and their permanent joints. All the effective factors can be subdivided into two categories: external (space) and internal (determined by the material properties) factors. Under the effect of these factors some physico-chemical processes causing degradation of operating parameters occur in materials, their permanent joints and coatings [4].

Table 1. Factors causing degradation of properties of structural materials and their permanent joints

| External factors | Internal factors |
|---|---|
| One-sided cyclic heating | Perfection of structure |
| Electromagnetic radiation | Impurities in materials |
| Bombardment by particles | Isotropy of properties |
| Vacuum | Density of dislocations |
| Aggressive residual atmosphere | Heterogeneity of materials |

| Processes in materials |
|---|
| Mechanical stresses |
| Changes in linear dimensions |
| Chemical and electrochemical corrosion |
| Microcracks |
| Evaporation of volatile components |
| Changes in structure |
| Phase transitions |
| Diffusion |

| |
|---|
| Changes in mechanical, optical, thermo- and electrophysical parameters |

The extent of the effect of these factors and processes depend on very many conditions and is little-studied as yet. Therefore, the level of degradation of properties by individual parameters is difficult to estimate. The scarce experimental data available and some theoretical background prove that the process of degradation of properties starts from the surface and undergoes at least two phases. First, the processes occur slowly and relatively uniformly, gradually penetrating the deeper layers. In the second phase there usually appear the local centres of the more intensive degradation, where from 10 to 100 % of the material service properties may be lost.

Lack of the experimental data necessitated the development of the purposeful comprehensive program of materials science studies.

The program is intended for the parallel mathematical and physical simulation of the space service conditions for fragments of large-sized structures made of various structural materials by welding, brazing, glueing and coating. Physical simulation will be realized by using the scale thin-walled models allowing to study both surface and depth processes of degradation at relatively short exposures. It is planned to perform exposure of the specimens both directly in space (to estimate the complex effect of all the factors) and in earth laboratories (to estimate them separately).

Realization of the program will allow to work out the system of criteria for well-founded selection of optimum structural materials, technology and methods to join them, the valid forecasting estimation of life of space structures, as well as to develop the procedure for their state current control.
The technological studies and the works performed on investigation of the materials science problems of welding in space are supposed to enrich the welding science and to stimulate the further progress of industrial welding technologies.

References

Paton, B.E., Lapchinskii, V.F., et al. (1989) Peculiarities of using Welding Processes for Fabrication and Repair of large-sized Structures in Space. Welding under extreme Conditions Proceedings of the International Institute of Welding, Helsinki, Pergamon Press, pp 181-188.

Ternovoj, E.G., Bondarev, A.A., et al. (1976) Study of some problems of weldability of aluminium alloys by using electron beam in zero-G, in Space studies in Ukraine, 6, 3-7, Naukova Dumka, Kiev.

Bondarev, A.A., Lapchinskii, V.F., et al. (1978) Study of structure and distribution of elements in welded joints made by electron beam in alloys 1201 and AMg6 under zero-G conditions, in Production and behaviour of materials in space, 21-29, Nauka, Moscow.

Paton, B.E. (1990) Technological problems of welding in space. Schwei en und Schneiden, 4...

Zagrebelny, A.A., Lukash, E.S., et al. (1988) Development of procedure for deposition of thin-film coatings under field conditions, in Proc. of XXII lectures dedicated to study of scientific heritage and to development of ideas of K.E. Tsyolkovskii, 81-95, IIET of the USSR Academy of Sciences, Moscow.